D1234884

Neural and Automata Networks

Mathematics and Its Applications

Managing Editor:

M. HAZEWINKEL
Centre for Mathematics and Computer Science, Amsterdam, The Netherlands

Editorial Board:

F. CALOGERO, *Università degli Studi di Roma, Italy*
Yu. I. MANIN, *Steklov Institute of Mathematics, Moscow, U.S.S.R.*
A. H. G. RINNOOY KAN, *Erasmus University, Rotterdam, The Netherlands*
G.-C. ROTA, *M.I.T., Cambridge, Mass., U.S.A.*

Volume 58

Neural and Automata Networks

Dynamical Behavior and Applications

by

Eric Goles

and

Servet Martínez

Departamento de Ingeniería Matemática,
Facultad de Ciencias Físicas y Matemáticas,
Universidad de Chile, Santiago, Chile

KLUWER ACADEMIC PUBLISHERS

DORDRECHT / BOSTON / LONDON

Sci
QA
267.5
C45
G65
1990

Library of Congress Cataloging in Publication Data

Goles, E.
    Neural and automata networks : dynamical behaviour and
applications / by Eric Goles Servet Martínez.
        p.   cm. -- (Mathematics and its applications)
    Includes bibliographical references.
    ISBN 0-7923-0632-5 (alk. paper)
    1. Cellular automata. 2. Computer networks. 3. Neural computers.
I. Martínez, Servet. II. Title. III. Series: Mathematics and its
applications (Kluwer Academic Publishers)
QA267.5.C45G65   1990
006.3--dc20   89-71622

ISBN 0–7923–0632–5

Published by Kluwer Academic Publishers,
P.O. Box 17, 3300 AA Dordrecht, The Netherlands.

Kluwer Academic Publishers incorporates
the publishing programmes of
D. Reidel, Martinus Nijhoff, Dr W. Junk and MTP Press.

Sold and distributed in the U.S.A. and Canada
by Kluwer Academic Publishers,
101 Philip Drive, Norwell, MA 02061, U.S.A.

In all other countries, sold and distributed
by Kluwer Academic Publishers Group,
P.O. Box 322, 3300 AH Dordrecht, The Netherlands.

Printed on acid-free paper

All Rights Reserved
© 1990 by Kluwer Academic Publishers
No part of the material protected by this copyright notice may be reproduced or
utilized in any form or by any means, electronic or mechanical
including photocopying, recording, or by any information storage and
retrieval system, without written permission from the copyright owner.

Printed in The Netherlands

ROBERT MANNING
STROZIER LIBRARY

JUN 22 1990

Tallahassee, Florida

To the memory of Moises Mellado

ROBERT MANNING
STROZIER LIBRARY

JUL 22 1991

Tallahassee, Florida

'Et moi, ..., si j'avait su comment en revenir, je n'y serais point allé.'

Jules Verne

The series is divergent; therefore we may be able to do something with it.

O. Heaviside

One service mathematics has rendered the human race. It has put common sense back where it belongs, on the topmost shelf next to the dusty canister labelled 'discarded nonsense'.

Eric T. Bell

Mathematics is a tool for thought. A highly necessary tool in a world where both feedback and nonlinearities abound. Similarly, all kinds of parts of mathematics serve as tools for other parts and for other sciences.

Applying a simple rewriting rule to the quote on the right above one finds such statements as: 'One service topology has rendered mathematical physics ...'; 'One service logic has rendered computer science ...'; 'One service category theory has rendered mathematics ...'. All arguably true. And all statements obtainable this way form part of the raison d'être of this series.

This series, *Mathematics and Its Applications*, started in 1977. Now that over one hundred volumes have appeared it seems opportune to reexamine its scope. At the time I wrote

"Growing specialization and diversification have brought a host of monographs and textbooks on increasingly specialized topics. However, the 'tree' of knowledge of mathematics and related fields does not grow only by putting forth new branches. It also happens, quite often in fact, that branches which were thought to be completely disparate are suddenly seen to be related. Further, the kind and level of sophistication of mathematics applied in various sciences has changed drastically in recent years: measure theory is used (non-trivially) in regional and theoretical economics; algebraic geometry interacts with physics; the Minkowsky lemma, coding theory and the structure of water meet one another in packing and covering theory; quantum fields, crystal defects and mathematical programming profit from homotopy theory; Lie algebras are relevant to filtering; and prediction and electrical engineering can use Stein spaces. And in addition to this there are such new emerging subdisciplines as 'experimental mathematics', 'CFD', 'completely integrable systems', 'chaos, synergetics and large-scale order', which are almost impossible to fit into the existing classification schemes. They draw upon widely different sections of mathematics."

By and large, all this still applies today. It is still true that at first sight mathematics seems rather fragmented and that to find, see, and exploit the deeper underlying interrelations more effort is needed and so are books that can help mathematicians and scientists do so. Accordingly MIA will continue to try to make such books available.

If anything, the description I gave in 1977 is now an understatement. To the examples of interaction areas one should add string theory where Riemann surfaces, algebraic geometry, modular functions, knots, quantum field theory, Kac-Moody algebras, monstrous moonshine (and more) all come together. And to the examples of things which can be usefully applied let me add the topic 'finite geometry'; a combination of words which sounds like it might not even exist, let alone be applicable. And yet it is being applied: to statistics via designs, to radar/sonar detection arrays (via finite projective planes), and to bus connections of VLSI chips (via difference sets). There seems to be no part of (so-called pure) mathematics that is not in immediate danger of being applied. And, accordingly, the applied mathematician needs to be aware of much more. Besides analysis and numerics, the traditional workhorses, he may need all kinds of combinatorics, algebra, probability, and so on.

In addition, the applied scientist needs to cope increasingly with the nonlinear world and the extra mathematical sophistication that this requires. For that is where the rewards are. Linear models are honest and a bit sad and depressing: proportional efforts and results. It is in the non-linear world that infinitesimal inputs may result in macroscopic outputs (or vice versa). To appreciate what I am hinting at: if electronics were linear we would have no fun with transistors and computers; we would have no TV; in fact you would not be reading these lines.

There is also no safety in ignoring such outlandish things as nonstandard analysis, superspace and anticommuting integration, p-adic and ultrametric space. All three have applications in both electrical engineering and physics. Once, complex numbers were equally outlandish, but they frequently proved the shortest path between 'real' results. Similarly, the first two topics named have already provided a number of 'wormhole' paths. There is no telling where all this is leading - fortunately.

Thus the original scope of the series, which for various (sound) reasons now comprises five sub-series: white (Japan), yellow (China), red (USSR), blue (Eastern Europe), and green (everything else), still applies. It has been enlarged a bit to include books treating of the tools from one subdiscipline which are used in others. Thus the series still aims at books dealing with:

- a central concept which plays an important role in several different mathematical and/or scientific specialization areas;
- new applications of the results and ideas from one area of scientific endeavour into another;
- influences which the results, problems and concepts of one field of enquiry have, and have had, on the development of another.

Neural networks are a hot topic at the moment. Just how overwhelmingly important they will turn out to be is still unclear, but their learning capabilities are such that an important place in the general scheme of things seems to be assured. These learning capabilities rest on the dynamic behaviour of neural networks and, more generally, networks of automata. The present book, which is largely based on the authors' own research, supplies a broad mathematical framework for studying such dynamical behaviour; it also presents applications to statistical physics.

The dynamics of networks such as studied in this volume are of interest to various groups of computer scientists, biologists, engineers, physicists, and mathematicians; all will find much here that is useful, stimulating, and applicable.

The shortest path between two truths in the real domain passes through the complex domain.

J. Hadamard

La physique ne nous donne pas seulement l'occasion de résoudre des problèmes ... elle nous fait pressentir la solution.

H. Poincaré

Never lend books, for no one ever returns them; the only books I have in my library are books that other folk have lent me.

Anatole France

The function of an expert is not to be more right than other people, but to be wrong for more sophisticated reasons.

David Butler

Bussum, January 1990   Michiel Hazewinkel

TABLE OF CONTENTS

FOREWORD

This book is intended to supply a broad mathematical framework for the study of the dynamics of Automata and Neural Networks. The main theoretical tools here developed are Lyapunov functionals which enable the description of limit orbits and the determination of bounds for transient lengths. Applications to models in Statistical Physics are made. We present relevant examples, illustrative simulations, and rigorous proofs of both classical and new results. We also provide a fairly complete bibliography on the subject.

The material in the book should appeal to mathematicians, physicists, biologists, computer science specialists, and engineers interested in Automata and Neural Networks, Dynamical Systems, Statistical Physics, Complexity and Optimization. Also scientists and engineers working in specifical applied models will find important material for their research work.

The book is mainly based upon the authors' research, an important part of which has been done in common, and on the collaboration of one of us (E.G.) with M. Cosnard (Ec. Norm. Sup. Lyon), F. Fogelman (U. Paris V), A.M. Odlyzko (Bell Labs.), J. Olivos (U. Chile), M. Tchuente (U. Cameroun), G. Vichniac (MIT).

Our approach to classical material, for instance algebraic invariants or Bounded Neural Networks, is entirely new and some of the proofs on this subjects are also original.

An important contribution to this book has been made by the members of our group on Automata Networks and Statistical Physics at the Universidad de Chile. In particular A. Maass furnished the examples on non-bounded periods of uniform 2-dimensional network (section 5.9), and M. Matamala develop the simulations of two-dimensional tie-rules given in Figures 3.3 and 3.4. Also we owe to C. Mejía (U. Paris V) the comparison between some continuous dynamics (see Figure 5.3).

The idea of writing this book was originally discussed with Enrique Tirapegui (U. Chile). We acknowledge his interest in this project.

We want to thank our sponsors and supporters whose concern and help were so important for this book. In particular we are indebted to Fondo Nacional de Ciencias of Chile, DTI of Universidad de Chile, Fundación Andes, French Cooperation at Chile and TWAS.

All our work has benefitted from the particular ambience at our Departamento de Ingeniería Matemática and our Facultad de Ciencias Físicas y Matemáticas de la Universidad de Chile, which is extremy positive toward research. We are indebted to our colleagues and to many people who have made these centers grow.

Mrs. Gladys Cavallone deserves a special mention for her very fine and hard work in typing this manuscript.

We also appreciate the patience of our families during the preparation of this book.

INTRODUCTION

1. Automata Networks

Automata Networks were introduced by Ulam, McCulloch and von Neumann to model phenomena studied in physics and biology [MP,U,VN1,VN2]. Automata Networks are discrete dynamical systems, in time and space. Roughly speaking, they are defined by a graph, either finite or infinite, where each site or vertex takes states in a finite set. Moreover, the state of a site changes according to a transition rule which takes into account only the state of its neighbors in the graph.

There exist several ways to introduce a dynamics in an Automata Network. The two most common ones are the synchronous (or parallel) and the sequential modes of updating the network. The former mode consists of bringing up to the next state, in a discrete time scale, all the sites simultaneously. The latter mode applies only in the case of finite graphs and consists of updating the sites one by one in a prescribed order. A comparison and theoretical results regarding these two updating modes are given in the context of discrete iterations in [Ro]. Other dynamics which we shall evoke briefly in the book are the memory and the block-sequential iteration. The former is similar to the synchronous updating, but it takes into account several previous configurations to compute the new state of the network. The latter is a combination of the synchronous and the sequential updating strategies. Within each block the evolution is synchronous, while between blocks it is sequential.

A particular case of Automata Networks are the Cellular Automata, which was the original model introduced by Ulam and von Neumann. In this case the graph is a regular lattice, usually $\mathbb{Z}^p$, the neighborhood structure and the transition rule are the same for all sites, the updating mode is synchronous.

Another important class of Automata Networks are the McCulloch-Pitts Automata, also called Neural Networks. In this model the graph is usually non-oriented and, at least in the earlier works, finite. The state set is binary, $\{0,1\}$ or $\{-1,1\}$ and the transition rule is a threshold function depending only on the neighbors and weighted by real numbers, i.e. $f(x_1,...,x_n) = \mathbb{1}\left(\sum_{j=1}^{n} a_j x_j - b\right)$ where $\mathbb{1}(u) = 1$ if $u \geq 0$ and 0 otherwise. In the case $\{-1,1\}$ the threshold is replaced by the sign function. Weights a_j are interpreted either as inhibitory synapses, if $a_j < 0$, or excitatory ones, if $a_j > 0$. The value b is the threshold. A Neural Network is a set of interconnected threshold or sign unit as above. A more realistic formal model of the nervous system taking inte account neuron refractary properties was proposed by Caianello [C2].

1

2. Complexity and Computing

From the beginning, the research in the field was oriented towards formalizing the complexity of automata as mesured by their computing capabilities. Since the paradigm of computing is the Turing Machine, several works deal with this model of abstract computation. For instance, the von Neumann Cellular Automaton has the capabilities of a Universal Turing Machine. In this automaton each site belongs to the two-dimensional lattice and is connected with four neighbors. For a particular initial configuration with a finite set of non-quiescent states divided in two parts called the "constructor" and the "organism", the synchronous evolution copies the "organism" in another part of the lattice similar to a reproduction phenomenon. The "constructor" may be seen as a decodification program of a Universal Turing Machine.

In the sixties Codd built a simpler automaton, also in a two dimensional lattice, with 5-neighbors and 8-states with the same capabilities of the von Neumann model [Co]. But the simplest one with the highest computing capabilities was given by Conway [BCG]: a 2-state, $\{0, 1\}$, 8-neighbor automaton in a two dimensional lattice, know as the "Game of Life", which may simulate a Universal Turing Machine and also has reproduction capabilities. Its local rule depends only on the state of the site and the sum of the states on the neighboorhood. A crucial point for this automaton to model the different units of a Turing Machine is the existence of gliders, which are finite configurations of 1's moving in the lattice.

Another way to determine computing capabilities of Cellular Automata was developped by Smith [Sm], who proved that automata defined in the $\mathbb{Z}$ lattice with next nearest interactions, are sufficient to simulate a Universal Turing Machine. He also presented a finer result: given a Turing Machine with n internal states and m alphabetic symbols, there exist a one-dimensional automaton with 6-neighbors and $\max(n, m) + 1$ states which simulates it. Since there exists a 4-symbol and 7-state Universal Turing Machine [Mi] he exhibits a one dimensional Cellular Automaton with 6-neighbors and 8 states.

In the Smith automaton only two sites may change their states in each synchronous iteration, so the parallel capabilities are wasted. This occurs in the majority of the previous simulations of Turing Machines. From this remark it is clear that the use of Cellular Automata as a model tool comes from applications where the parallel treatment and the local interaction are relevant. This occurs in several problems in physics and biology as well as in image processing and filter algorithms.

Concerning Neural Networks, at the beginning the interest comes from their complex computing capabilities. In fact, this allows to assert that the real nervous system, which inspired Neural Networks as a very simplified model, has as least the powerful capabilities exhibited by the formal model. In this optic, McCulloch

and Pitts showed that a finite Neural Network computes any logical function [MP].
Also, it was easily shown that it simulates any Sequential Machine [Mi] and if an
infinite number of sites is provided, Turing Machines. Also important is the work
of Kleene, who characterized the formal languages recognized by Neural Networks
as the Regular ones [K].

3. Earlier Dynamical Results

In all the above models, their dynamical behaviour is impredictable. In fact
there do not exist short-cut theorems serving to establish the asymptotic dynamics.
The only way to know the evolution is to simulate the automaton in a computer
and to watch it evolve. But as we shall show further in the book, dynamical results
for some classes of Automata Networks can be obtained.

In this context, one of the first studies was done for Cellular Automata with
q states identified as the integers $(\mod q)$ and a local linear rule on this set. It
was shown by Waksman [Wa] and Winograd [Wi] that for q prime, every finite
configuration reproduces itself indefinitely during the synchronous dynamics. Per-
haps this helps to understand some of the typical class-2 patters of the Wolfram
classification [W2] which present self-similarity or fractal dimensions.

For a Cellular Automata model of a reaction-diffusion equation, Gremberg,
Greene and Hastings give some combinatorial tools to study the asymptotic dynamics [GGH]. They give conditions under which waves travelling indefinitely in
the cellular space or configurations which converge to the quiescent state occur.
More recently, the periodic structure of the steady state was studied in [AR].

Concerning Neural Network dynamics, we must refer to results obtained by
Kitagawa, Shingai, Kobuchi and Tchuente for some one- and two-dimensional
regular lattices and particular threshold local rules. Kitagawa was one of the first
authors to discuss the dynamics of several bounded Neural Networks [Ki1]. For a
rectangle with 4-neighbourhood structure, excitatory local weights and a threshold
equal to 3, he characterized the set of fixed points and showed the convergence of
the sequential update. Kobuchi studied two-dimensional Neural Networks with the
von Neumann neighbourhood and a local threshold rule with positive threshold.
He found the following generic behaviour: given the initial condition where all the
sites are in the quiescent state except a corner site, only four kinds of patterns in the
asymptotic dynamics can exist [Ko]. Shingai studied the periodic behaviour of one-
and two-dimensional Bounded Uniform Networks and gave results on the periods
of their synchronous updating. Roughly, according to the vector of weights he ob-
serves periods dependent or independent of the network size [Sh1,Sh2]. Tchuente
studied the majority binary local rule, which is a particular threshold function, in
one- and two-dimensional lattices. In the one-dimensional case, he showed that for

arbitrary connected neighborhoods the majority rule converges to fixed points [T1] and in two dimensions with threshold equal to 2 and the von Neumann interaction he characterized the dynamics as the convergence to fixed points or two-cycles. Moreover, he characterized the distribution of states in the steady phase [T3].

All these results are extremely dependent on the particular structure of each case, so it is not possible to generalize the results to other Neural or Automata Networks.

Besides, entirely different approaches exist, coming from other theories. A pioneer work was done by Little who related Neural Networks to the spin glass problem [Li,LiS]. His work has been reprised in recent years by several physicists, as we will explain further on. Concerning finite Markov fields and ergodic theory, the dynamics of some related Cellular Automata have been studied [DKT,KS,W1]. For instance probabilistic Cellular Automata and Erasing Automata (i.e. deterministic automata whose dynamics erases any finite configuration). Toom [To,ToM] and Galperin [Ga1,Ga2] have given characterizations for the erasing property and proved that the recognition of Erasing Automata is undecidable. Experimental results on automata dynamics have played an important role in the development of the subject. The first one consist of the simulation of cristal growth conducted by Ulam [U]. Another systematic experimentation was done by Kauffman to model genetic interactions. He observed the synchronous dynamics of random boolean networks. In these, each site has a boolean transition function where the neighbours and the function are chosen at random at the begining. The principal result obtained in the simulation of this model was that periods and number of cycles vary as $n^{1/2}$, where n is the number of sites in the network [Ka]. Kauffman networks have been also studied in [AFS,DP,DW,F,FG1].

4. Some Current Trends on Automata Networks

A renewed interes in Automata Networks has come very recently from different areas and situations: the new parallel architectures for computers (for instance hypercubes, connection machines, etc), the development of non expensive hardware boards to simulate automata evolutions [TfM], new learning algorithms in a Neural Network enviroment, and deep relationship between the physics of disordered systems and Automata Networks.

A few years ago learning algorithms appeared in order to categorize classes of patterns through Neural Networks whose interconnection weights are learned from examples and with good generalizations capabilities. Two very powerful and today well known algorithms are the Boltzman Machine and the Gradient Back Propagation for Multilayer Neural Networks [RM]. In both cases the limitations of the first neural learning model, the perceptron [MiP], disappears and several important hard problems have been approached with good performances. In both cases the most used local rule, the sigmoid rule, is slighly different from the threshold one. It will be studied in Chapter 5 as a particular case of cyclically monotone functions. It is to be noted that these algorithms, as well as other variations, have had an enormous impact on research and applications. Nowadays there exist several journals dedicated to Neural Computing.

The relation to physics stems from two cardinal works: Wolfram's studies on Cellular Automata [W1] and the Hopfield model of Associative Memories [Ho2]. The former consist of massive simulations of one-dimensional Cellular Automata, with their interpretation in the framework of statistical physics. Wolfram proposes four complexity classes for automata according to the asymptotic pattern generated by the synchronous dynamics. Also, he posed several problems, some of which are entirely new while some others are reinterpretations of older ones. Mainly, they deal with the cross-over among dynamical systems, statistical mechanics, computing and language complexity theory [W3]. This program is currently in progress and some interesting results may be seen for instance in [BFW,CY,DGT,GM5,Hu,Pe]. Also Wolfram motivated a lot of people to use Automata Networks as a paradigm of complex systems and also to use them to model some physical problems [BM,V1,V2]. In this context, the lattice Gas Hydrodynamics is one of the most noted examples of approaches to physical models by Automata Networks. From the precursory works of Hardy, de Pazzis and Pomeau [HP1,HP2] lattice gas models have been developed in theory and practice (see for instance the proceedings of the workshop on large nonlinear systems dedicated to this approach [FH]).

Hopfield's work consists of modelling Associative Memories by a Neural Network. Given a set of boolean vectors to be memorized, he defines a threshold network whose symmetric weights are defined by the Hebb correlation rule between the patterns. He proved that such patterns are fixed points of the networks and, by using the analogy with the spin glass problem, he further showed that the sequential network dynamics is driven by a Lyapunov operator, as does the spin glass Hamiltonian in the related problem [Ho2]. His findings drove a lot of physicists to develop more theoretical results in the framework of statistical mechanics [AGS,D,FW,MPV,Per].

On the other hand, a rigorous mathematical framework for studing the dynamics of some classes of Automata Networks has been developed in the last years (see references of the introduction an in particular [BFW,CG,DGT,DKT,FRT,G3, Pe,Ro,T1]). An important part of these works will be presented in this book.

5. The Book

This book furnishes a mathematical framework which allows an explanation for small periods in the limit behaviour of a large class of Automata Networks. This class includes symmetrical Neural Networks which will be treated in detail in several chapters. The tools here introduced also give optimal bounds of transient lengths. The main hypotheses we use are discussed in detail and several examples and simulations illustrate different classes of dynamics.

Although our techniques are in principle build for classical Neural Networks with state set $\{0,1\}$ or $\{-1,1\}$, we are able, by using convex analysis theory, to study evolution of continuous and higher dimensional networks.

But our approach also permits explanations for other phenomena arising in statistical physics. We present two noted examples: the Thermodynamical limit of Gibbs states in the Bethe lattice and a class of automata inspired in the Potts model.

The results presented in this book have been obtained in the last few years, most of them after 1979. Some of them are unpublished. For each major result we give the reference where it appeared for the first time, even if its original version is more particular. Usually the proof we furnish is slighly different from the original one. The global framework of the book is entirely new in that it unifies a series of results which have appeared in entirely different contexts.

In Chapter 1 we develop complexity theory of Automata Networks. While the theory is classical, its consequences in automata dynamics are still not well-known. We supply some classical results which evince why it is not possible to study dynamics of Automata Networks with a unified approach. In fact without no restrictions a diversified and extremely rich dynamics can appear.

In Chapter 2 we impose a crucial hypothesis: symmetry of weight connections. This condition is fulfilled by a large class of interesting Neural Networks, in particular by those studied in the Hopfield model and most of the earlier results presented in section 3 of the book. Symmetry implies a very simple asymptotic behaviour. The periods of the steady state are ≤ 2. We prove it by using the algebraic invariants introduced in [GO1,GO3,GO4] which capture the combinatorics of the period of the orbits. This methodology is succefully developed in the study of several dynamics which include synchronous, sequential, block-sequential and memory updating of Neural Networks as well as evolution of Majority Networks.

In Chapter 3 Lyapunov functionals for Neural Networks are introduced. After the pionner work of Hopfield [Ho2], this theory was mainly developed in [FGM,G3,GFP]. These functionals do not only explain that periods are ≤ 2 but also give very rich information on transient lengths. We lay emphasis on the optimility of these bounds by developing in detail the synchronous iteration with integer connection weights. For sequential updating the Lyapunov functional also helps to describe transient behaviour.

The antisymmetrical case is also analyzed with these techniques. In this context we determine cycles of period 4 for the synchronous update and period 2 for the sequential case. We finally develop a class of symmetric Neural Networks whose behaviour illustrates a surprising phenomenon: while the steady states of symmetrical networks are simple, their transient lengths can be very complicated, in fact they can be exponential in the size of the network.

In Chapter 4 we study evolutions of one and two dimensional Uniform Neural Networks. We develop in detail two-state Majority Networks and give the form of their fixed points and their transient behaviour. But the chapter is mostly dedicated to furnish an original proof of Shingai's theorem which asserts that a one dimensional Uniform Bounded Neural Network with next nearest interactions and fixed boundary conditions has period ≤ 4. This result is cited in [Sh2] but the original paper, [Sh1], is extremely difficult to translate rigorously. Apparently his proof consists of an exhaustive analysis of cases. We are also led to study several cases, but a significant simplification can be achieved when a sign hypothesis is fulfilled in fact it implies that the network can be symmetrized, hence results obtained in previous chapters are applied. The longest analysis is performed when we study the cases which correspond to period $T = 3$. For the multidimensional case the sign hypothesis also implies $T \leq 2$ and when the hypothesis does not hold we can exhibit configurations with non-bounded periods.

In Chapter 5 we introduce a wide class of networks that contains as a particular case the Neural Networks. We begin by solving the continuous case, i.e. the bridge between threshold and increasing local rules. We approach continuous functions from different points of view but in all of them we search for an extension of Lyapunov functionals obtained in Chapter 3. First we introduce positive functions for which the extension is direct, we describe this class and exhibit some multi-state majorities and some continuous rules. Then we approach the continuous case as a limit of Multithreshold Neural Networks. But the most general case, which corresponds to Cyclically Monotone Networks, is studied by means of convex analysis techniques. This approach has been developed in [GM2,GM4] and independently in [PT1,PT2], in the framework of algebraic invariants, and in [PE] in the context of convex optimization.

In this context, we study the synchronous and sequential updating modes and develop generalizations to make the matrices and functions present in our dynamics larger. Related optimization problems are discussed and we present the sequential iteration with a continuous rule as a hill-climbing strategy for quadratic discrete combinatorial optimization problems.

The methodology developed in Chapter 5 can also cover problems arising in classical statistical physics. In Chapter 6 we discuss the thermodinamics limit in the Bethe lattice. There we use the particular form that the distribution of thermodynamic limits take, in fact their equations can be read as a dynamics of a continuous positive network. As a problem closely related to this one, we also study the asymptotic dynamics of non-linear evolutions of probability vectors generalizing the usual Markov chains [GM4].

In Chapter 7 we take a slighly different point of view. By taking generalizations of the classical Hamiltonian of the Potts model we associate local rules that make this decrease when they are applied in sequential mode. In fact, this dynamics leads to fixed points which are local minima of the Hamiltonian. Nevertheless, when these compatible rules are updated synchronously the dynamics may be extremely complex. In particular it can simulate a Universal Turing Machine. In spite of this negative result, there exist some particular classes of compatible functions whose synchronous iteration may be analyzed with Lyapunov operators derived from the Potts Hamiltonian. As particular cases we find still the multistate majority rule and some local smoothing rules used in image processing. In this context we study in detail the convergence of the phase-unwrapping algorithm to determine the phase of a complex function or to obtain global average in digital images [GOd,OR].

This book can be read in several ways, depending on the interests of the reader.

Chapter 1 deals with the fundamental concepts of Automata and Neural Networks and it might interest most of the people related to this subject. This is also the case for Chapter 4, which analyzed classical Uniform Neural Networks by means of usual combinatorial arguments.

For researchers the core of the book is Chapter 3 because Lyapunov functionals, which are the most succesful tools for studying Neural Networks, are introduced therein. In developing the chapter major dynamical results are obtained.

Applied mathematicians can be specially interested in Chapters 2 and 5. The former provides the algebraic tools while the latter gives a general framework, as well as discusses relations with optimization problems.

Physicists can find in Chapters 6 and 7 important models from statistical physics analyzed by means of the tools developed in Chapter 3 and 5.

The notation is introduced in Chapter 1 and in this sense its reading is essential for understandying the rest of the book.

The following diagram depicts the relations between chapters:

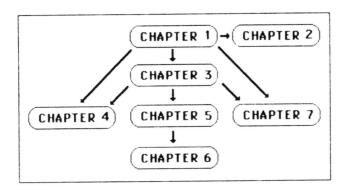

Here and at the end of each chapter we have included the corresponding list of references. They have been collected and augmented in References at the end of the book.

References

[AFS] Atlan, H., F. Fogelman-Soulie, J. Salomon, G. Weisbuch, *Random Boolean Networks*, Cybernetics and Systems, 12, 1981, 103.

[AGS] Amit, J.D., H. Gutfreund, M. Sompolinsky, *Spin-Glass Models of Neural Networks*, Phys. Review A, 32(2), 1985, 1007-1018.

[AR] Allouche, J.P., Ch. Reder, *Oscillations Spatio-Temporelles Engendrees par un Automate Cellulaire*, Disc. Applied Maths., 1984, 215-254.

[BCG] Berlekamp, E.R., J.H. Conway, R.K. Guy, *Winning Ways*, Ac. Press, 1985, 2, Chapter 25.

[BFW] Bienenstock, E., F. Fogelman-Soulie, G. Weisbuch (eds), Disordered Systems and Biological Organization, NATO ASI Series F: Computer and Systems Sciences, 20, 1986.

[BM] Bidaux, R., P. Manneville (eds), *Proc. Workshop "Cellular Automata and Modeling of Complex Physical Systems"*, Les Houches, February 1989, to appear in Springer-Verlag.

[C2] Caianiello, E.R., *Outline of a Theory of Thought-Processes and Thinking Machines*, J. Theor. Biol., 2, 1961, 204-235.

[CG] Cosnard, M., E. Goles, *Dynamique d'un Automate a Mémoire Modélisant le Fonctionement d'un Neurone*, C.R.A.S., 299(10), 1984, 459-461.

[Co] Codd, E.F., *Cellular Automata*, Acad. Press, 1988.

[CY] Culik, K., S. Yu, *Undecidability of C.A. Classification Schemes*, Complex Systems, 2(2), 1988, 177-190.

[D] Derrida, B., *Dynamics of Automata, Spin Glasses and Neural Network Models*, Preprint, Service de Physique Théorique, CEN-Saclay, France, 1987.

[DGT] Demongeot, J., E. Goles, M. Tchuente (eds), *Dynamical Systems and Cellular Automata*, Acad. Press, 1985.

[DKT] Dobrushin, R.L., V.I. Kryukov, A.L. Toom, *Locally Interacting Systems and their Application in Biology*, Lecture Notes in Mathematics, N⁰653, 1978.

[DP] Derrida, B., Y. Pomeau, *Random Networks of Automata: A simple Annealed Approximation*, Europhysics Letters, 1(2), 1986, 45-49.

[DW] Derrida, B., G. Weisbuch, *Evolution of Overlaps between Configurations in Random Boolean Networks*, J. Physique, 47, 1986, 1297-1303.

[F] Fogelman-Soulie, F., *Contributions a une Theorie du Calcul sur Reseaux*, Thesis, IMAG, Grenoble, France, 1985.

[FG1] Fogelman-Soulie, F., E. Goles, G. Weisbuch, *Specific Roles of the Different Boolean Mappings in Random Networks*, Bull. Math. Biol., 44(5), 1982, 715-730.

[FGM] Fogelman-Soulié, F., E. Goles, S. Martínez, C. Mejía, *Energy Functions in Neural Networks with Continuous Local Functions*, 1988, submitted Complex Systems.

[FH] Frisch, U., B. Hasslacher, S. Orszag, S. Wolfram, *Proc. of "Workshop on Large Nonlinear Systems"*, 1986, Complex Systems, 1(4), 1987.

[FRT] Fogelman-Soulie, F., Y. Robert, M. Tchuente (eds), *Automata Networks in Computer Science; Theory and Applications*, Nonlinear Science Series, Manchester Univ. Press, 1987.

[FW] Fogelman-Soulie, F., G. Weisbuch, *Random Iterations of Threshold Networks and Associative Memory*, SIAM J. on Computing, 16, 1987, 203-220.

[G3] Goles, E., *Comportement Dynamique de Reseaux d'Automates*, Thesis, IMAG, Grenoble, 1985.

[Ga1] Galperin, G.A., *One-dimensional Automata Networks with Monotonic Local Interactions*, Problemy Peredachi Informatsii, 12(4), 1976, 74-87.

[Ga2] Galperin, G.A., *One-Dimensional Monotonic Tesselations with Memory* in Locally Interacting Systems and their Application in Biology, R.L. Dobrushin et al (eds), Lecture Notes in Mathematics, N⁰653, 1978, 56-71.

[GFP] Goles, E., F. Fogelman-Soulie, D. Pellegrin, *The Energy as a Tool for the Study of Threshold Networks*, Disc. App. Math., 12, 1985, 261-277.

[GGH] Greenberg, J.M., C. Greene, S.P. Hastings *A Combinatorial Problem Arising in the Study of Reaction-Diffusion Equations*, SIAM J. Algebraic and Discrete Meths, 1, 1980, 34-42.

[GM2] Goles, E., S. Martínez, *Properties of Positive Functions and the Dynamics of Associated Automata Networks*, Discrete Appl. Math. 18, 1987, 39-46.

[GM4] Goles, E., S. Martínez, *The One-Site Distributions of Gibbs States on Bethe Lattice are Probability Vectors of Period ≤ 2 for a Nonlinear Transformation*, J. Stat. Physics, 52(1/2), 1988, 267-285.

[GM5] Goles, E., S. Martínez (eds), *Proc. Congrés Franco-Chilien en Math. Appliquées*, 1986, in Revista de Matemáticas Aplicadas, 9(2), 1988.

[GO1] Goles, E., J. Olivos, *Compartement Itératif des Fonctions a Multiseuil*, Information and Control, 45(3), 1980, 300-313.

[GO3] Goles, E., J. Olivos, *Compartement Periodique des Fonctions a Seuil Binaires et Applications*, Disc. Appl. Maths., 3, 1981, 93-105.

[GO4] Goles, E., J. Olivos, *Periodic Behaviour of Generalized Threshold Functions*, Disc. Maths., 30, 1980, 187-189.

[GOd] Goles, E., A.M. Odlyzko, *Decreasing Energy Functions and Lengths of Transients for some Lengths of Transients for Some Cellular Automata*, Complex Systems, 2(5), 1988, 501-507.

[Ho2] Hopfield, J.H., *Neural Networks and Physical Systems with Emergent Collective Computational Abilities*, Proc. Nat. Acad. Sci., UAS, 79, 1982, 2554-2558.

[HP1] Hardy, J., O. de Pazzis, Y. Pomeau, *Time Evolution of a Two-Dimensional Model System: Invariant States and Time Correlation Functions*, J. Math. Phy., 14, 1973, 174.

[HP2] Hardy, J., O. de Pazzis, Y. Pomeau, *Molecular Dynamics of a Classical Lattice Gas: Transport Properties and Time Correlation Functions*, Phys. Rev. A. 13, 1976, 1949.

[Hu] Hurd, L.P., *Formal Language Characterizations of Cellular Automaton Limit Sets*, Complex Systems, 1, 1987, 69-80.

[K] Kleene, S.C., *Representation of Events in Nerve Nets and Finite Automata* in Automata Studies, C.E. Shannon and J. McCarthy (eds), Annals of Mathematics Studies, 34, Princeton Univ. Press, 1956, 3-41.

[Ka] Kauffman, S.A., *Behaviour of Randomly Constructed Genetic Nets* in Towards a Theoretical Biology, C.H. Waddington (ed), 3, Edinburgh Univ. Press, 1970, 18-46.

[Ki1] Kitagawa, T., *Cell Space Approaches in Biomathematics*, Math. Biosciences, 19, 1974, 27-71.

[Ko] Kobuchi, Y., *Signal Propagation in 2-Dimensional Threshold Cellular Space*, J. of Math. Biol., 3, 1976, 297-312.

[KS] Kindermann, R., J.L. Sneel, *Markov Random Fields and their Applications*, Series on Contemporary Mathematics, AMS, 1, 1980.

[Li] Little, W.A. *Existence of Persistent States in the Brain*, Math. Bios, 19, 1974, 101.

[LiS] Little, W.A., G.L. Shaw, *Analytic Study of the Memory Storage Capacity of a Neural Network*, Math. Bios, 39, 1978, 281-290.

[Mi] Minsky, M.L., *Computation: Finite and Infinite Machines*, Prentice-Hall series in Automatic Computation, 1967.

[MiP] Minsky, M., S. Papert, *Perceptrons, an Introduction to Computational Geometry*, MIT Press, 1969.

[MP] McCulloch, W., W. Pitts, *A Logical Calculus of the Ideas Immanent in Nervous Activity*, Bull. Math. Biophysics, 5, 1943, 115-133.

[MPV] Mezard, M., G. Parisi, M.A. Virasoro (eds), *Spin Glass Theory and Beyond*, Lecture Notes in Physics, 9, World Scientific, 1987.

[OR] Odlyzko, A.M., D.J. Randall, *On the Periods of Some Graph Transformations*, Complex Systems, 1, 1987, 203-210.

[PE] Pham Dinh Tao, S. El Bernoussi, *Iterative Behaviour, Fixed Point of a Class of Monotone Operators. Application to Non-Symmetric Threshold Functions*, Disc. Maths., 70, 1988, 85-101.

[Pe] Peliti, L. (ed), *Disordered Systems and Biological Models*, Procc of the Workshop and Disordered Systems and Biol. Modelling, Bogotá Colombia, 1987, World Scientific, CIF Series, 14, 1989.

[Per] Peretto, P., *Collective Properties of Neural Networks: A Statistical Physics Approach*, Biol. Cybern., 50, 1984, 51-62.

[PT1] Poljak, S., D. Turzik, *On Pre-Periods of Discrete Influence Systems*, Disc. Appl. Maths., 13, 1986, 33-39.

[PT2] Poljak, S., D. Turzik, *On an Application of Convexity to Discrete Systems*, Disc. Appl. Math., 13, 1986, 27-32.

[RM] Rumelhart, D.E., J.L. McClelland (eds), *Parallel and Distributed Processing: Explorations in the Microstructure of Cognition*, MIT Press, 1986.

[Ro] Robert, F., *Discrete Iterations. A Metric Study*, Springer Series in Computational Mathematics, Sprnger-Verlag, 1986.

[RoT] Robert, Y., M. Tchuente, *Connection-Graph and Iteration-Graph of Monotone Boolean Functions*, Disc. Appl. Maths., 11,1985, 245-253.

[Sh1] Shingai, R., *Maximum Period of 2-Dimensional Uniform Neural Networks*, Inf. and Control, 41, 1979, 324-341.

[Sh2] Shingai, R., *The Maximum Period Realized in 1-D Uniform Neural Networks*, Trans. IECE, Japan, E61, 1978, 804-808.

[Sm] Smith, A.R., *Simple Computation-Universal Cellular Spaces*, J. ACM, 18(3), 1971, 339-353.

[T1] Tchuente, M., *Contribution a l'Etude des Methodes de Calcul pour des Systemes de Type Cooperatif*, Thesis, IMAG, Grenoble, France, 1982.

[T3] Tchuente, M., *Evolution de Certains Automates Cellulaires Uniformes Binaires A Seuil*, Seminaire 265, IMAG, Grenoble, 1977.

[TfM] Toffoli, T., M. Margolus, *Cellular Automata Machines: A New Environment for Modeling*, MIT Press, 1987.

[To] Toom, A.L., *Monotonic Binary Cellular Automata*, Problemy Peredaci Informacii, 12(1), 1976, 48-54.

[ToM] Toom, A.L., L.G. Mityushin, *Two Results Regarding Noncomputability for Univariate Cellular Automata*, Problemy Peredaci Informacii, 12(2), 1976, 69-75.

[U] Ulam S., *On Some Mathematical Problems Connected with Patterns of Growth of Figures* in Essays on Cellular Automata, A.W. Burks (ed), Univ. of Illinois Press, 1970, 219-243.

[V1] Vichniac, G., *Simulating Physics with Cellular Automata*, Physica 10D, 1984, 96-116.

[V2] Vichniac G., *Cellular Automata Models of Disordered and Organization* in Disordered Systems ans Biol. Org., E. Bienenstock et al (eds), NATO ASI Series F, 20, 1986, 3-19.

[VN1] von Neumann, J., *Theory of Self-Reproducing Automata*, A. W. Burks (ed), Univ. of Illinois Press, 1966.

[VN2] Von Neumann, J., *The General and Logical Theory of Automata* in Hixon Synposium Proc., 1948 in J.N. Neumann Collected Works, A.H. Taub (ed), Pergamon Press, V,288-328, 1963.

[W1] WolframS., *Theory and Applications of Cellular Automata*, World Scientific, 1986.

[W2] Wolfram, S., *Universality and Complexity in Cellular Automata*, Physica 10D, 1984, 1-35.

[W3] Wolfram, S., *Twenty Problems in the Theory of Cellular Automata*, Phys. Scripta T9, 1985, 170.

[Wa] Waksman, A., *A Model of Replication*, J.A.C.M., 16(1), 1966, 178-188.

[Wi] Winograd, T., *A Simple Algorithm for Self-Reproduction*, MIT, Project MAC, Artificial Intelligence, Memo 198, 1970.

1. AUTOMATA NETWORKS

1.1. Introduction

In this chapter we introduce several examples of Automata Networks and we briefly give some complexity results for this kind of models from a computer science standpoint, namely results on computing capabilities of Automata Networks and, as a particular case, of Neural Networks. The complexity results have been obtained in a more general framework by several authors (see for instance [Gr,K,Mi,Sm,T1]) but they are not still well known in the context of automata dynamics. Another reason to present this class of results is that they necessarily imply that some restrictions on the automaton must be imposed in order that its dynamics can be described. In fact, Automata Networks are extremely complex computing machines and in most cases the only way to get information on its dynamical behaviour is by computer simulations; there are no short cut theorems. Hence we are led to take up some hypotheses, the most fruitful and less restrictive ones being those concerning the symmetry of connections of the Automata Network. In fact they allow the characterization of the steady state and of the length of the transient for the asymptotic dynamics of a wide class of automata which include the symmetric Neural Networks.

1.2. Definitions Regarding Automata Networks

Let I be a set of sites or vertices which is not necessarily finite. An *Automata Network* defined on I is a triple $\mathcal{A} = (G, Q, (f_i : i \in I))$ where:

- $G = (I, V)$ is a graph on I with connections given by the set $V \subset I \times I$. We assume G to be locally finite, which means that each neighbourhood $V_i = \{j \in I : (j, i) \in V\}$ is finite, i.e., $|V_i| < \infty$ for any $i \in I$. As a system of neighbourhoods, $(V_i : i \in I)$ also determines the set of connections by the equality $V = \{(j, i) : j \in V_i, i \in I\}$; hence we can also denote the graph by $G = (I, (V_i : i \in I))$.

- Q is the set of states, which in most of the cases is assumed to be finite.

- $f_i : Q^{|V_i|} \to Q$ is the transition function associated to vertex i. The automaton's global transition function $F : Q^I \to Q^I$, defined on the set of configurations Q^I, is constructed with the local transition functions $(f_i : i \in I)$ and with some kind of updating rule, for instance a synchronous or a sequential one.

When the graph is finite, i.e. $|I| < \infty$, there is no loss of generality if we assume that each f_i is defined from $Q^{|I|}$ into Q.

To visualize a graph, an arc $(j, i) \in V$ is drawn as an arrow from j to i: $j \to i$. For instance in the graph $G = (I, V)$ in Figure 1.1 below, the set of sites

15

is $I = \{1,2,3,4\}$ and the collection of neighbourhoods consists of $V_1 = \{2,3,4\}$, $V_2 = \{1\}$, $V_3 = \{2\}$, and $V_4 = \{3\}$

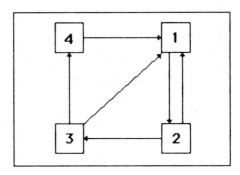

Figure 1.1. Four vertex oriented graph.

A graph $G = (I, V)$ is called non oriented if the equivalence $(j, i) \in V$ iff $(i, j) \in V$ (or, $j \in V_i$ iff $i \in V_j$) holds. In most of the book we shall deal with this kind of graphs. In such graphs the arcs (j, i), (i, j) are drawn as a simple link between i and j without an arrow head.

The updating rule of an automaton $\mathcal{A}$ can take different forms. In this chapter we only consider the sequential and the synchronous (or parallel) types.

The *synchronous* iteration of a network $\mathcal{A} = (G, Q, (f_i : i \in I))$ results when all the sites of the networks are updated at the same time, so its dynamics is given by the equation

$$x(t+1) = F_{\mathcal{A}}(x(t)), \text{ whose } i\text{-th component } x_i(t+1) = (F_{\mathcal{A}}(x(t)))_i \text{ is}$$
$$x_i(t+1) = f_i(x_j(t) : j \in V_i) \tag{1.1}$$

In a *sequential* iteration the sites are updated one at a time in a prescribed order, given by a relation $\leq$ in I. Thus the dynamics of the sequential iteration is:

$$x_i(t+1) = f_i(y_j : j \in V_i)$$

$$\text{where} \quad y_j = \begin{cases} x_j(t+1) & \text{if } j < i \\ x_j(t) & \text{if } j \geq i \end{cases} \tag{1.2}$$

Sequential updating is only used in the case of finite graphs. Thus, for $I = \{1, ..., n\}$ endowed with its canonical order $\leq$ (if not so we relabel the sites) and assuming $f_i : Q^{|I|} \to Q$, the evolution equation (1.2) of sequential updating takes the form:

$$x_1(t+1) = f_1(x_1(t), ..., x_n(t))$$
$$x_i(t+1) = f_i(x_1(t+1), ..., x_{i-1}(t+1), x_i(t), ..., x_n(t)) \quad \text{for } 1 < i \leq n$$

Sequential updating was introduced in Numerical Analysis to solve linear systems and it is known as the Gauss-Seidel relaxation. In the context of Automata it is used for instance in associative memory models [Ho2] and in some combinatorial optimization problems [HoT].

Other updating models that we shall briefly evoke in this book are:

- the block-sequential iteration, in which the space of sites is partitioned into blocks $I = \bigcup_{k \in K} I_k$. Within each block I_k the updating is synchronous and among the different classes $(I_k : k \in K)$ it is performed in a sequential form following some prescribed order $\leq$ in K;

- the k-memory step iteration, in which each iteration depends on the k previous states of the Automata Network.

It is important to note that for some automata the sequential and synchronous updating may have completely different dynamical behaviours. Nevertheless, it is easy to see that both modes always have the same fixed points. Precise results on different kinds of discrete iterations and relationships between them can be seen in [Ro].

1.3. Cellular Automata.

Let $I = \mathbb{Z}^d$. If a set of connections $V \subset \mathbb{Z}^d \times \mathbb{Z}^d$ is translation invariant, i.e. $(j,i) \in V$ iff $(j+k, i+k) \in V$, the graph $G = (\mathbb{Z}^d, V)$ is called a cellular space. In terms of the neighbourhoods the translation invariant condition is expressed by equalities, namely: $V_i = (i - j) + V_j$ for any $i, j \in \mathbb{Z}^d$. From $V_i = i + V_0$, we deduce that all the information of the neighbour structure is contained in V_0, so we write $G = (\mathbb{Z}^d, V_0)$. The graph G is non-oriented iff V_0 is symmetric around the origin: $V_0 = -V_0$.

For $I = \mathbb{Z}$ the neighbourhood V_0 is usually taken as $V_0 = \{-q, ..., 0, ..., p\}$ for some $p, q > 0$, hence $V_i = i + V_0 = \{i - q, ..., i, ..., i + p\}$.

For $I = \mathbb{Z}^2$ the commonest choices for V_0 are the following:

$V_0^N = \{(1,0), (-1,0), (0,0), (0,1), (0,-1)\}$ the von Neumann neighbourhood

and $V_0^M = V_0^N \cup \{(1,1), (1,-1), (-1,1), (-1,-1)\}$ the Moore neighbourhood

Hence, for $i = (i_1, i_2)$ the von Neumann neighbourhood is $V_i^N = i + V_0^N = \{(i_1+1, i_2), (i_1-1, i_2), (i_1, i_2), (i_1, i_2+1), (i_1, i_2-1)\}$ and the Moore neighbourhood is $V_i^N \cup \{(i_1+1, i_2+1), (i_1+1, i_2-1), (i_1-1, i_2+1), (i_1-1, i_2-1)\}$. Graphically we have:

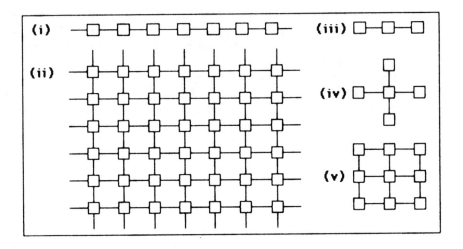

Figure 1.2. (i) $1 - D$ Cellular space. (ii) $2 - D$ Cellular space. (iii) Neighbourhood $\{-1, 0, 1\}$. (iv) von Neumann neighbourhood. (v) Moore neighbourhood.

Cellular Automata are automata defined on cellular spaces whose transition function is also translation invariant: $f_i = f$ for any $i \in \mathbb{Z}^d$ with $f : Q^{|V_0|} \to Q$. We assume that the finite set of states Q contains a particular state noted 0, the quiescent state: $f(0, ..., 0) = 0$.

Hence the synchronous evolution of a Cellular Automata $\mathcal{A} = (\mathbb{Z}^d, V_0, Q, f)$ is given by :

$$x(t + 1) = F_{\mathcal{A}}(x(t)), \text{ which } i\text{-th component } x_i(t+1) = (F_{\mathcal{A}}(x(t)))_i \text{ is:}$$
$$x_i(t + 1) = f(x_j(t) : j \in i + V_0) \tag{1.3}$$

For this class of automata it is important to study the evolution of initial finite configurations. More precisely, for every $x \in Q^{\mathbb{Z}^d}$ we define its support as the set of sites whose state is not the quiescent one:

$$\text{supp}(x) = \{i \in \mathbb{Z}^d : x_i \neq 0\}$$

We say x is finite if $|\text{supp}(x)| < \infty$ and we denote the collection of such x's by $Q^{\mathbb{Z}^d}_{fin}$.

Since $f(0, ..., 0) = 0$, the automata acts on the set of finite configurations, $F_{\mathcal{A}} : Q^{\mathbb{Z}^d}_{fin} \to Q^{\mathbb{Z}^d}_{fin}$. The set of configurations, $\mathcal{C}_{\mathcal{A}}$, of a Cellular Automata will consist of the finite configurations: $\mathcal{C}_{\mathcal{A}} = Q^{\mathbb{Z}^d}_{fin}$.

There are also finite Cellular Automata defined on the torus $\mathbb{Z}^d_{\bar{T}}$. Let $\bar{T} = (T_1, ..., T_d)$ be a sequence of strictly positive integers and mod $\bar{T}$ be the equivalence relation it induces on $\mathbb{Z}^d$: $(x_1,...,x_d) = (x'_1,...,x'_d)(\bmod \bar{T})$ iff $x_j - x'_j = k_j \bar{T}_j$ for some $k_j \in \mathbb{Z}$, $j = 1,...,d$. The torus $\mathbb{Z}^d_{\bar{T}}$ is the set of equivalence classes $\mathbb{Z}^d / \bmod \bar{T}$ which is an additive group with the addition operation. The condition of invariance on the set of neighbourhoods means that $V_i = i + V_0(\bmod \bar{T})$. The usual 0-neighbourhoods V_0 are $V_0 = \{-1,0,1\}$, $V_0 = \{-1,1\}$, $V_0 = \mathbb{Z}^d_{\bar{T}}$, all of which satisfy the invariant property.

In order that the finite cellular space $(\mathbb{Z}^d_{\bar{T}}, V_0)$ can support a Cellular Automata we also impose that the local functions are translational invariants, hence a finite Cellular Automata is a tuple $\mathcal{A} = (\mathbb{Z}^d_{\bar{T}}, V_0, Q, f)$ where $f : Q^{|V_0|} \to Q$ is the local function. Its synchronous evolution is also given by $x(t+1) = F_{\mathcal{A}}(x(t))$ with $(F_{\mathcal{A}}(x))_i = f(x_j : j \in i + V_0(\bmod \bar{T}))$.

Concerning the evolutions defined on infinite cellular spaces $(\mathbb{Z}^d, V_0)$ by local functions, i.e. by functions which depend only on a few number of sites of the network, we can distinguish two main problems. The first is to describe the global dynamics $x(t) = F_{\mathcal{A}}^t(x)$ of the Cellular Automata solely on the knowledge of the lattice structure and the local function. The second is to characterize the set of global dynamics F which come from local functions.

The latter problem was solved by Hedlund [He] and generalized by Richardson [Ri]. They characterize this class of global functions F as consisting of those which are continuous with respect to the product topology on $Q^{\mathbb{Z}^d}$ (we impose the discret topology on Q) and which commute with the shift σ, i.e. satisfy $F \circ \sigma = \sigma \circ F$, where $\sigma((x_i)_{i \in \mathbb{Z}^d}) = ((x_{i+\bar{1}})_{i \in \mathbb{Z}^d})$.

As for the former problem, it has no general answer. As we shall see in the next paragraph, Cellular Automata may simulate Universal Turing Machines; hence it is not possible to give short cut theorems characterizing the asymptotic dynamics of the automata without imposing some hypothesis on the local function. Obviously, when the structure of the Automata Network is more general (not cellular) the situation is more difficult to analyse.

1.4. Complexity Results for Automata Networks

As pointed out in the previous section, the dynamical behaviour of an automaton may be broadly described as complex. To define a precise criterium which captures the idea of complexity, several concepts have been introduced, some of them as chaos or non-ergodicity [L,LN,Ml,W1], borrowed from the theory of classical dynamical systems. A more specific approach to automata complexity has been followed up by Wolfram [W2] and, on the basis of an enormous experimental work, he has proposed a classification of one-dimensional Cellular Automata (1-D C.A.) in four classes:

(1) : 1-D C.A. where any initial condition converges to some fixed point.
(2) : 1-D C.A. where the limits of initial conditions are cycles.
(3) : 1-D C.A. which accepts fractal patterns and arbitrary periods.
(4) : 1-D C.A. with breaking symmetry configurations (as gliders) and long-lived localized patterns.

The above classification is empirical and difficult to apply. In this view, one negative result is the following: given an arbitrary 1-D C.A. it is undecidable whether it belongs to class (1) [ToM]. Other undecidability results of this kind may be seen in [CY].

We can also define complexity based on a computer science approach, a C.A. being complex if it performs non-trivial algorithms or, more precisely, if it can simulate a Universal Turing Machine. In connection with this problem, there exists a conjecture of Wolfram concerning particular C.A. of Class-4 which are suspected to simulate Universal Turing Machines [W2].

In this section we present two results which make it evident that, automata networks, in spite of their simplicity, have powerful computing capabilities. To make our notions precise, let us first introduce Sequential and Turing Machines.

A *finite Sequential Machine* is a tuple $S M = (Q, \Sigma, \delta)$ where:

- Q is a finite set of internal states with two distinguished states: q_0 the initial one, and q_f the final one.

- $\delta : Q \times \Sigma \to Q$ is the transition function which associates to every pair (q, s) consisting of an input and an internal state a new internal state $\delta(q, s)$.

The Sequential Machine works as follows: it starts in the initial state q_i and reads an input symbol $s \in \Sigma$; on doing so it changes its internal state to $\delta(q, s)$. A word $\bar{s} = s_0, ..., s_k$ of the language of this machine $S M$ is a finite sequence of symbols of Σ which is a path from q_i to q_f i.e. $\delta(q^{(j)}, s_j) = q^{(j+1)}$ with $q^{(0)} = q_i$, $q^{(k+1)} = q_f$.

Kleene [K] proved that the Sequential Machine languages are in fact the Regular Languages (i.e. sets of words built by union, concatenation and closure of elements in the alphabet Σ); this means that the language of any Sequential Machine is Regular and that for any Regular Language there exists a Sequential Machine which realizes it. This is a complexity result in the sense that the computing capabilities of $S M$ are characterized by Regular Language recognition. More complex languages need more powerful computing machines. Results relating automata networks dynamics with languages can be found in [CY,Hu,L,W1].

The finiteness of a Sequential Machine implies that there exist algorithms which cannot be implemented regardless of the data size. For instance, the multiplication of two arbitrary integers depends on their size, and a Sequential Machine needs a number of internal states varying with the size of the numbers. In order to overcome this problem, we give the more general model of computation: the Turing Machine.

A *Turing Machine* is a tuple $\mathcal{T} = (Q, \Sigma, \delta)$ where:

- Q is a finite set of internal states, with a distinguished state q_0 which is the initial state.

- Σ is the finite set of tape symbols used for input and output, with a particular element called the blank and noted 0.

- $\delta : Q \times \Sigma \rightarrow Q \times \Sigma \times \{-1, 0, 1\}$ is the transition function which associates to every pair (q, s), consisting of an internal state and a scanned symbol, a vector $\delta(q, s) = (q', s', \ell)$ with q' being the new internal state, s' the symbol printed in the place of s, and ℓ the moving direction: $\ell = -1$ codes the movement of the machine's head to the left, $\ell = 1$ to the right, and $\ell = 0$ means it remains at the same place.

A Turing Machine may be visualized as a head containing a Sequential Machine, with an external environment given by a doubly infinite tape. Initially on the tape there exists a finite word $\bar{s}^0$ of symbols belonging to $\Sigma \setminus \{0\}$, all other symbols being 0, and the head is in the internal state q_0 and points to the leftmost element of the input word. Call m_0 the coordinate of the tape where the leftmost element of $\bar{s}^0$ lies, suppose this word is of length p, note by $s^0_{m_0}, ..., s^0_{m_0+p-1}$ the word of symbols belonging to $\Sigma \setminus \{0\}$ and note $s^0_i = 0$ for $i \neq m_0, ..., m_0 + p - 1$. If $\delta(q_0, s^0_{m_0}) = (q', s', \ell)$ then in the next step of time the internal state of $\mathcal{T}$ is q', the new symbol in coordinate m_0 is s', and the head moves to position $m_0 + \ell$.

At each time step the configuration of the Turing Machine $\mathcal{T}$ is determined by: its internal state $q \in Q$, the tape coordinate $m \in \mathbb{Z}$ at which the head points, and the string of symbols $\bar{s} = (s_i : i \in \mathbb{Z}) \in \Sigma^{\mathbb{Z}}_{fin}$ on the tape, where $\Sigma^{\mathbb{Z}}_{fin}$ contains the countable set of elements $\bar{s} = (s_i : i \in \mathbb{Z}) \in \Sigma^{\mathbb{Z}}$ with only a finite number of symbols different from 0.

The evolution of $\mathcal{T}$ can be viewed as an action on the set of configurations $\mathcal{C}_{\mathcal{T}} = Q \times \mathbb{Z} \times \Sigma^{\mathbb{Z}}_{fin}$. Let $F_{\mathcal{T}} : \mathcal{C}_{\mathcal{T}} \rightarrow \mathcal{C}_{\mathcal{T}}$ be this evolution, then when we apply $F_{\mathcal{T}}$ to the configuration $(q, m, (s_i : i \in \mathbb{Z})) \in Q \times \mathbb{Z} \times \Sigma^{\mathbb{Z}}_{fin}$ we get:

$$F_{\mathcal{T}}(q, m, (s_i : i \in \mathbb{Z})) = (q', m', (s'_i : i \in \mathbb{Z}))$$
$$\text{where } \delta(q, s_m) = (q', s'_m, m' - m) \text{ and } s'_i = s_i \text{ if } i \neq m$$

(recall that $m' \in \{m - 1, m, m + 1\}$).

The initial condition is $q = q_0$, $m = m_0$, $s_i = s^0_i$ for $i \in \mathbb{Z}$. Hence at step $t \geq 0$ the Turing Machine has evolved to the configuration $F^t_{\mathcal{T}}(q_0, m_0, (s^0_i : i \in \mathbb{Z}))$. Suppose there exists $\tau \geq 0$ such that the machine halts at this step:

$$F^{\tau+1}_{\mathcal{T}}(q_0, m_0, (s^0_i : i \in \mathbb{Z})) = F^{\tau}_{\mathcal{T}}(q_0, m_0, (s^0_i : i \in \mathbb{Z}))$$

Call $(\tilde{q}, \tilde{m}, (\tilde{s}_i : i \in \mathbb{Z}))$ the above configuration; then the result of the computation is the finite word $\tilde{s}_{p_1}, ..., \tilde{s}_{p_2}$ written on the tape at this step τ (the coordinates $p_1 \leq p_2$ are such that $s_i = 0$ for $i < p_1$ or $i > p_2$).

It should be noted that the "halt problem" of a Turing Machine is undecidable [Mi].

From the definition of a Turing Machine $T = (Q, \Sigma, \delta)$ it is clear that it can be coded as a finite vector $w_T = (q, s, \delta(q, s) = (q', s', \ell) : q \in Q, s \in \Sigma)$.

Furthermore, it is possible to build a Turing Machine $T_u = (Q_u, \Sigma_u, \delta_u)$ which is called *Universal* because it satisfies the following property: for any Turing Machine T it is possible to take a coded version of w_T on the tape of T_u such that for any initial condition $(q_0, \tilde{s}_0)$ of T the machine T_u realizes the computing of T.

Moreover there exists Universal Turing Machines $T_u = (Q_u, \Sigma_u, \delta_u)$ with Q_u containing 7 internal states, and an alphabet Σ_u of 4 symbols [Mi].

Rigorous complexity results deal with the following definition. An Automata Network A is said to simulate a Turing Machine T if there exists a one-to-one function φ (the encoding function) from the set of configurations of T in the set of finite configurations of A commuting with the evolutions, that is $\varphi : C_T \to C_A$ satisfies:

$$\varphi \circ F_T = F_A \circ \varphi$$

where F_T gives the evolution of T and F_A the global evolution of A.

An Automata Network A is said to be Universal if it simulates a Universal Turing Machine.

Theorem 1.1. [Sm] Any Turing Machine $\tau = (Q, \Sigma, \delta)$ can be simulated by a one dimensional Cellular Automata $A = (\mathbb{Z}, V_0 = \{-1, 0, 1\}, \tilde{Q}, f)$.

Proof. Take $\tilde{Q} = ((Q \cup \{*\}) \times \Sigma)$ where $*$ is a symbol not belonging to Q. Recall that the set of configurations of the Cellular Automata A is:

$$C_A = \tilde{Q}_{fin}^{\mathbb{Z}} = \{x \in \tilde{Q}^{\mathbb{Z}} : \text{supp } x \text{ is finite }\}$$

and that the set of configurations of the Turing Machine is $C_T = Q \times \mathbb{Z} \times Q_{fin}^{\mathbb{Z}}$. Consider the following encoding function φ, which is clearly one-to-one:

$$\varphi : C_T \to C_A \quad \text{is such that} \quad \varphi(q, m, (s_i : i \in \mathbb{Z})) = (\tilde{q}_i : i \in \mathbb{Z})$$
$$\tilde{q}_m = (q, s_m) \quad \text{and} \quad q_i = (*, s_i) \quad \text{for} \quad i \neq m.$$

On the Cellular Automaton $A = (\mathbb{Z}, V_0 = \{-1, 0, 1\}, \tilde{Q}, f)$ we take the following local function $f : \tilde{Q}^3 \to \tilde{Q}$.

$$f((q, s), (*, x), (*, y)) = \begin{cases} (q', x) & \text{if } \delta(q, s) = (q', s', 1) \text{ for some } s' \in \Sigma \\ (*, x) & \text{otherwise} \end{cases}$$

$$f((*,x),(q,s),(*,y)) = \begin{cases} (q',s') \text{ if } \delta(q,s) = (q',s',0) \\ (*,s') \text{ if } \delta(q,s) = (q',s',\ell) \text{ for } \ell \in \{-1,1\} \text{ and } q' \in Q \end{cases}$$

$$f((*,x),(*,y),(q,s)) = \begin{cases} (q',y) \text{ if } \delta(q,s) = (q',s',-1) \text{ for some } s' \in \Sigma \\ (*,y) \text{ otherwise} \end{cases}$$

For any other configuration $(a,b,c) \in \tilde{Q}^3$ we take $f(a,b,c) = b$.
It is direct to show that $\varphi \circ F_T = F_A \circ \varphi$ ∎

The Cellular Automaton $A = (\mathbb{Z}, \{-1,0,1\}, \tilde{Q}, f)$ previously constructed simulates each step of the Turing Machine only by activating two sites of the automaton. Hence it is a sequential simulation and the parallel possibilities of a Cellular Automaton are wasted.

On the other hand the state set $\tilde{Q}$ has cardinality $|\tilde{Q}| = (|Q|+1) \times |\Sigma|$. A more performant simulation with a neighbourhood of 6 elements may be obtained, whose state set satisfies $|\tilde{Q}| = \max(|Q|, |\Sigma|) + 1$ (see [Sm]).

Since there exist Universal Turing Machines, from the previous Theorem we have:

Corollary 1.1. There exist Universal Cellular Automata. ∎

Since we can build Universal Turing Machines with 7 states and 4 symbols, the above remark implies that there exists a Universal Cellular Automata with 8 states and a neighbourhood of size 6.

1.5. Neural Networks

Introduced by McCulloch and Pitts [MP] to model some features of the nervous system, Neural Networks are able to simulate any Sequential Machine or Turing Machine if an infinite number of cells is provided. Furthermore, this kind of automata have had an enormous development for applications in several domains: physics, artificial intelligence, and biology [BFW,Ho2,HoT,RM].

We shall present some complexity results which show that some regularities on the connections of the network must be assumed in order that a dynamical study of Neural Networks is possible. In fact, without any restrictive hypothesis Neural Networks can exhibit a dynamical behaviour as complex as that of Turing Machines.

A Neural Network N is a particular class of automata network. Its graph $G = (I, V)$ posseses a weighted structure. Thus if V is the set of connections, to every arc $(j,i) \in V$ we associate a real number $a_{ij} \in \mathbb{R}$ which represents its weight. If $(j,i) \notin V$ we put $a_{ij} = 0$. The graph $G = (I,V)$ being locally finite, the structure of the matrix of interactions A is such that $a_{ij} = 0$ for all $j \in I$ except for a finite set $V_i \subset I$. The state set is $Q = \{0,1\}$ or $Q = \{-1,+1\}$. In this book we shall usually take $Q = \{0,1\}$, the state $y = 0$ means inhibited and $y = 1$ means excited. The local functions are the following:

$$f_i : \{0,1\}^{|V_i|} \to \{0,1\}, \quad f_i(x_j : j \in V_i) = \mathbb{1}\Big(\sum_{j \in V_i} a_{ij} x_j - b_i\Big) \tag{1.4}$$

where $\mathbb{1}$ is the threshold function:

$$\mathbb{1}(u) = \begin{cases} 1 \text{ if } u \geq 0 \\ 0 \text{ if } u < 0 \end{cases} \tag{1.5}$$

This evolution is interpreted as follows. The quantity $a_{ij} > 0$ means an excitatory synaptic weight and $a_{ij} < 0$ an inhibitory weight, hence the state of the neuron at cell i will be excited iff the weighted sum $\sum_{j \in V_i} a_{ij} x_j$ is greater than its threshold b_i.

The matrix $A = (a_{ij} : i,j \in I)$ is called the matrix of connections and $b = (b_i : i \in I)$ is called the threshold vector. A *Neural Network* $\mathcal{N}$ will be noted by the tuple $\mathcal{N} = (I,V,A,b)$. As the neighbour structure can also be read in the interaction matrix A, we often write $\mathcal{N} = (I,A,b)$. Its synchronous evolution is given by the equation:

$$x_i(t+1) = \mathbb{1}\Big(\sum_{j \in I} a_{ij} x_j(t) - b_i\Big), \quad i \in I, \ t \geq 0 \tag{1.6}$$

Let $\bar{\mathbb{1}}(u_i : i \in I) = (\mathbb{1}(u_i) : i \in I)$ be the multidimensional threshold function. Define the following transformation from $\{0,1\}^{|I|}$ into itself: $F_{\mathcal{N}}(x) = \bar{\mathbb{1}}(Ax - b)$. Hence the synchronous dynamics of a Neural Network can be written:

$$x(t+1) = F_{\mathcal{N}}(x(t)) = \bar{\mathbb{1}}(Ax(t) - b) \quad \text{for } t \geq 0 \tag{1.7}$$

Recall that the locally finiteness property of the graph $G = (I,V)$ implies that for any $x \in \{0,1\}^{|I|}$ the point Ax is well defined for any set I, whether the set is finite or not.

The sequential evolution of a finite network with $I = \{1,...,n\}$ endowed with the canonical order is:

$$x_i(t+1) = \mathbb{1}\Big(\sum_{j<i} a_{ij} x_j(t+1) + \sum_{j \geq i} a_{ij} x_j(t) - b_i\Big), \ i \in I \tag{1.8}$$

When the state set Q is $\{-1, +1\}$ the local function of the Neural Network is replaced by $\text{sign}\,(\sum_{j \in V_i} a_{ij} x_j - b_i)$ where $\text{sign}\,(u) = 1$ if $u \geq 0$, -1 if $u < 0$. Obviously both kinds of Neural Networks (with $\mathbb{1}$ and sign) are entirely equivalent. It is important to point out that the linearity of the argument at each unit $(\sum_{j \in V_i} a_{ij} x_j - b_i)$ will be essential for the understanding of its dynamical behaviour.

In order to study the complexity of Neural Networks let us first show that any finite application can be conveniently coded by finite Neural Networks.

Lemma 1.1. Let D_1, D_2 be finite sets and $h : D_1 \to D_2$ some function. Then there exist a finite Neural Network $\mathcal{N}^{(0)} = (I^{(0)}, A^{(0)}, b^{(0)})$ and two encoding functions:

$$\Psi_\ell : D_\ell \to \{0,1\}^{|I^{(0)}|} \quad \ell = 1, 2$$

such that:

$$\Psi_2 \circ h = F_{\mathcal{N}^{(0)}} \circ \psi_1 \quad \text{where } F_{\mathcal{N}^{(0)}}(x) = \bar{\mathbb{1}}(A^{(0)} x - b^{(0)}).$$

When $D_1 = D_2$ the above construction also satisfies $\Psi_1 = \Psi_2$.

Proof. We can restrict ourselves to show the lemma for sets of the form $D_1 = \{1, ..., m_1\}$, $D_2 = \{m_1 - m, ..., n_1 + n_2 - m\}$, where $n_\ell = |D_\ell|$ for $\ell = 1, 2$ and $m = |D_1 \cap D_2|$.

Take $I^{(0)} = D_1 \cup D_2$, so $|I^{(0)}| = n_1 + n_2 - m$. For $k \in D_\ell$ pick $\Psi_i(k) = x$, where $x_i = 1$ if $i = k$, $x_i = 0$ if $i \neq k$. Recall $(\Psi_1)_{D_1 \cap D_2} = (\Psi_2)_{D_1 \cap D_2}$.

Now define:
$$a_{ij}^{(0)} = \begin{cases} 1 \text{ if } j \in D_1, \, i \in D_2, \, i = h(j). \\ 0 \text{ otherwise} \end{cases}$$

We take the threshold vector $b_i^{(0)} = \frac{1}{2}$ for $i \in I$.

Let $k \in D_1$ and $x = \Psi_1(k)$, then:

$$\sum_{j \in I} a_{ij}^{(0)} x_j - b_i^{(0)} = \begin{cases} \frac{1}{2} \text{ if } i = h(j) \\ 0 \text{ otherwise} \end{cases}$$

Hence $\Psi_2(h(k)) = \bar{\mathbb{1}}(A^{(0)} \Psi_1(k) - b^{(0)})$ and the lemma is shown. ∎

In the above scheme Ψ_1 is the encoding of the input and Ψ_2 of the output. In order to get a dynamics we must compose Ψ_2^{-1} with $F_{\mathcal{N}^{(0)}} \circ \Psi_1$. Thus we obtain $h^t = (\Psi_2^{-1} \circ F_{\mathcal{N}^{(0)}} \circ \Psi_1)^t$. When $D_1 = D_2$ we get $\Psi_1 = \Psi_2 = \Psi$ so $h^t = \Psi^{-1} \circ F_{\mathcal{N}^{(0)}}^t \circ \Psi$, and in this case all the dynamical properties of h are contained in those of $F_{\mathcal{N}^{(0)}}$.

We shall apply the above lemma for local functions, i.e. for functions of the type $f : Q^{|V_0|} \to Q$. If $|V_0| = 1$ we need only one encoding function Ψ and in the case $|V_0| \geq 2$ the sets $Q^{|V_0|}$ and Q are disjoint, so Ψ_1 and Ψ_2 activate entirely different subsets of cells of the Neural Network $\mathcal{N}^{(0)}$.

Similar results on simulation of Sequential Machines by Neural Networks and their computing capabilities can be seen in [Mi].

Proposition 1.1. For any Cellular Automata $\mathcal{A} = (\mathbb{Z}^d, V_0, Q, f)$ there exists a Neural Network $\mathcal{N} = (I, A, b)$ and two encoding functions, $\varphi_\ell : C_\mathcal{A} \to C_\mathcal{N}$ for $\ell = 1, 2$, such that:

$$\varphi_2 \circ F_\mathcal{A} = F_\mathcal{N} \circ \varphi_1$$

The set of cells I is countably infinite.

Proof. Let $f : Q^{|V_0|} \to Q$ be the local function. Associate to it a network $\mathcal{N}^{(0)} = (I^{(0)}, A^{(0)}, b^{(0)})$ and a couple of functions Ψ_1, Ψ_2 given by the above lemma $\Psi_2 \circ f = F_{\mathcal{N}^{(0)}} \circ \Psi_1$.

Take $I = \bigcup_{k \in \mathbb{Z}^d} I^{(k)}$ where $I^{(k)}$ is just a copy of $I^{(0)}$ in site k. We define the matrix A by:

$$a_{ij} = \begin{cases} a_{ij}^{(0)} & \text{if } i, j \in I^{(k)} \text{ for some } k \\ 0 & \text{otherwise} \end{cases}$$

and the threshold vector equal to the constant $\frac{1}{2}$.

For a configuration $x = (x_k \in \mathbb{Z}^d) \in C_\mathcal{A}$ define:

$$\varphi_1(x) = (\Psi_1(x_r : r \in V_k) : k \in \mathbb{Z}^d) \text{ and } \varphi_2(x) = (\Psi_2(x_k) : k \in \mathbb{Z}^d).$$

Since x has a finite support the element $A\varphi_1(x) - b$ is well defined, its k-block being $A^{(0)}\Psi_1(x_r : r \in V_k) - b^{(0)}$. So $\bar{\mathbb{1}}(A^{(0)}\Psi_1(x_r : r \in V_k) - b^{(0)}) = \Psi_2(f(x_r : x \in V_k))$. We have shown $\varphi_2 \circ F_\mathcal{A} = F_\mathcal{N} \circ \varphi_1$. ∎

A similar construction for Universal Cellular Automata reveals that Neural Networks can have a highly complex behaviour. Then regularity assumptions must be made on $\mathcal{N} = (I, A, b)$, even in the finite case, in order that its dynamics can be described.

1.6. Examples of Automata Networks

In this section we present some examples of the evolution of automata, basically in one and two dimensional celular spaces. The majority of the examples are related to our theoretical results and some of them will be treated in more detail in subsequent chapters. Other examples may be seen in [W1].

1.6.1. XOR Networks. Let $\mathcal{A} = (G, Q = \{0,1\}, (f_i : i \in I))$ be a finite automaton where G is the graph of Fig 1.1 and the local functions are: $f_1(x_2, x_3, x_4) = x_2 + x_3 + x_4 \pmod 2$, $f_2(x_1) = x_1, f_3(x_2) = x_2, f_4(x_3) = x_3$ for $x = (x_1, x_2, x_3, x_4) \in \{0,1\}^4$.

The synchronous iteration associated to $\mathcal{A}$ is:

$$x_1(t+1) = x_2(t) + x_3(t) + x_4(t) \pmod 2, \quad x_2(t+1) = x_1(t),$$
$$x_3(t+1) = x_2(t), \quad x_4(t+1) = x_3(t) \quad \text{for } t \geq 0, \ x(0) \in \{0,1\}^4$$

The iteration graph for all the vertices of the 4-hypercube is given in Figure 1.3.i. Each vector is represented by the integer associated to the modulo two representation.

Figures 1.3.ii and 1.3.iii show the iteration graphs of the sequential update associates to the orders $4 < 3 < 2 < 1$ and $1 < 2 < 3 < 4$ respectively. Clearly 0 and 15 are fixed points which are invariant under different updates.

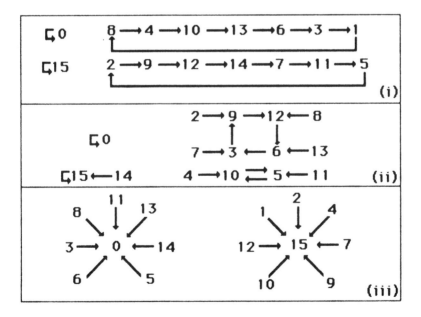

Figure 1.3. (i) Synchronous update. (ii) Sequential update in order $4 < 3 < 2 < 1$. (iii) Sequential update in order $1 < 2 < 3 < 4$.

1.6.2. Next Majority Rule. Let $\mathcal{A} = (\mathbb{Z}_{200}, V_0 = \{-1, 0, 1\}, Q = \{0, 1, 2, 3\}, f)$ be a one-dimensional automaton in a 200 torus. The local function is:

$$f(x_{-1}, x_0, x_1) = \begin{cases} x_0 + 1 (\text{mod} 4) & \text{if} \\ \quad |\{j \in V_0 / x_j = x_0 + 1 (\text{mod} 4)\}| \geq |\{j \in V_0 / x_j = x_0\}| \\ x_0 \text{ otherwise} \end{cases}$$

This automaton updated in synchronous mode admits configurations breaking the space symmetry (gliders) as well as non-bounded cycles. Furthermore, as we shall see in Chapter 7, it may compute any logic function by coding binary information as gliders. Examples of its dynamical behaviour are given in Figure 1.4.

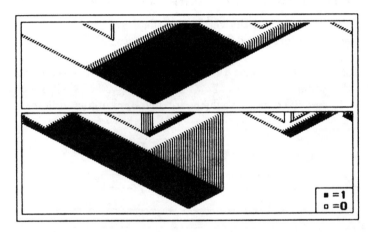

Figure 1.4. Synchronous dynamics of the next majority rule in a 200-torus. The vertical axis represents the discrete time steps.

1.6.3. Multithreshold Automaton. Let $\mathcal{A} = (\mathbb{Z}_{200}, V_0 = \{-2, -1, 0, 1, 2\}, Q = \{0, 1, 2, 3\})$ be the one-dimensional automaton in a 200 torus with the local function:

$$f(x_{-2}, x_{-1}, x_0, x_1, x_2) = \begin{cases} 0 & \text{if} \quad \sum_{j \in V_0} x_j < b_0 \\ 1 & \text{if} \quad b_0 \leq \sum_{j \in V_0} x_j < b_1 \\ 2 & \text{if} \quad b_1 \leq \sum_{j \in V_0} x_j < b_2 \\ 3 & \text{otherwise} \end{cases}$$

This class of functions will be studied, as a particular case of cyclically monotone applications, in Chapter 5. Multithreshold Automata generalize Neural Networks, which have only one threshold per site.

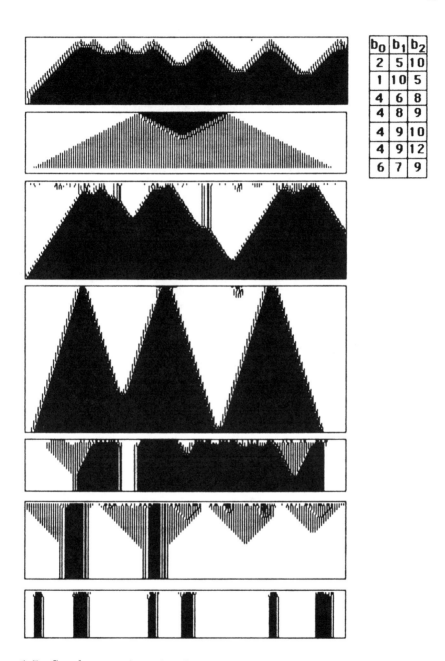

Figure 1.5. Synchronous iteration for Multithreshold Automata. The different images depend on the threshold vector (b_0, b_1, b_2).

Figure 1.5 shows patterns generated by the synchronous iteration for different threshold vectors $b = (b_0, b_1, b_2) \in I\!N^3$. It is easy to see that for higher values of $S = \sum_i b_i$ the finite initial configurations grow indefinitely, otherwise they converge to fixed points and for small values of S to the quiescent state. In the latter case A is called Erasing Automata. In this context given a Multithreshold Automata A with $Q = \{0, 1, 2, 3\}$, $V_0 = \{-h, ..., 0, ..., +h\}$, and a vector b satisfying the following regularity conditions:

$$b_0 \in [h + 2, 2h + 1[, \quad b_1 \in [2h + 3, 4h + 2[,$$

$$b_2 \in [4h + 3, 5h + 3[, \quad b_0 + b_1 \in [4h + 5, +\infty[,$$

it is proved in [GT2] that $S \leq 9h + 8$ implies that A is not erasing, and that $S \geq 9h + 10$ implies that A is erasing. For the critical case $S = 9h + 9$ the two behaviours can appear.

The erasing problem is a hard one to deal with. It was shown in [ToM] that, in the restricted class of monotone local rules, it is undecidable. Other results about Erasing Automata may be seen in [Ga2].

1.6.4. The Ising Automaton. Let $\{I_0, I_1\}$ be a partition of the set of sites $I = \{0, ..., 199\}$. Consider the following one-dimensional automaton $A = (\mathbb{Z}_{200}, V_0 = \{-2, -1, 0, 1, 2\}, Q = \{0, 1\}, (f_i : i \in I))$:

$$f_i(x_{i-2}, x_{i-1}, x_i, x_{i+1}, x_{i+2}) = \begin{cases} 0 & \text{if} \quad \sum_{j \in V_0 + i} x_j - 2 < 0 \\ x_i & \text{if} \quad i \in I_0 \text{ and } \sum_{j \in V_0 + i} x_j - 2 = 0 \\ 1 - x_i & \text{if} \quad i \in I_1 \text{ and } \sum_{j \in V_0 + i} x_j - 2 = 0 \\ 1 & \text{otherwise} \end{cases}$$

Local functions are majority rules with a different tie-break: in case of tie, a site belonging to I_0 remains in its previous value; but a site belonging to I_1 flips its state.

In Figure 1.6 we exhibit dynamical patterns only for the sequential iterations in the order $0 < 1 < 2 < ... < 199$. As we shall see in Chapter 3, the synchronous iteration always converges to fixed points or two-cycles. The patterns generated by the sequential update depend on the parameter $p = |I_0|/200$, which is the density of "frozen sites". For $p = 0$ (no freeze dynamics) we remark that any initial configuration is a left-shift, for $0 < p < 1$ we obtain a periodic behaviour, and for $p = 1$ (the freeze case), only fixed points. Some other results, in higher dimensions, will be presented in Chapter 3.

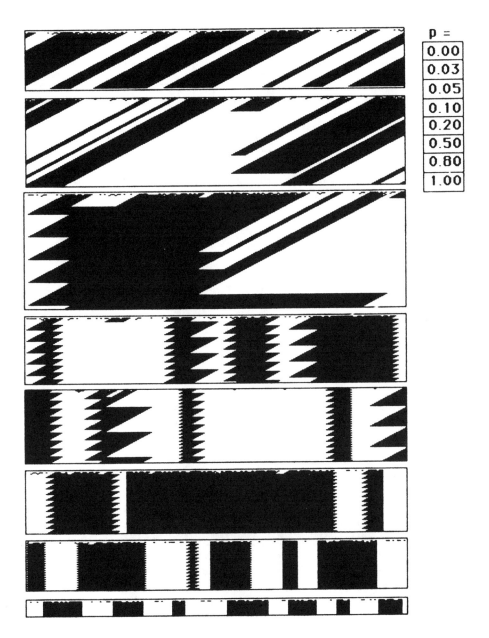

p =
| 0.00 |
| 0.03 |
| 0.05 |
| 0.10 |
| 0.20 |
| 0.50 |
| 0.80 |
| 1.00 |

Figure 1.6. Sequential dynamics of the Ising Automaton in a 200 torus. Different images depend on the density, $p = |I_0|/200$, of frozen sites.

1.6.5. Bounded Neural Network (BNN). Consider $\mathcal{A} = (I \times I, Q = \{0, 1\}, f)$, for $I = \{1, ..., n\}$ and the following neighbourhood:

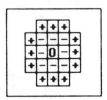

Figure 1.7. Neighbourhood, $V_{(0,0)}$, of the BNN. The internal sites are weighted with -1 and the external ones with $+1$.

The local function is

$$
f_1\left(x_{(i,i')}; (i, i') \in V_{(0,0)}\right) = \begin{cases} 0 & \text{if} \quad \sum_{j \in V_0} a_j x_j < 0 \\ 1 - x_{(0,0)} & \text{if} \quad \sum_{j \in V_0} a_j x_j = 0 \\ 1 & \text{otherwise} \end{cases}
$$

where $a_j = -1$ for internal sites and $a_j = +1$ otherwise.

Any site $(j, j') \in \mathbb{Z} \times \mathbb{Z} \setminus I \times I$ is fixed in the quiescent state 0 (boundary condition).

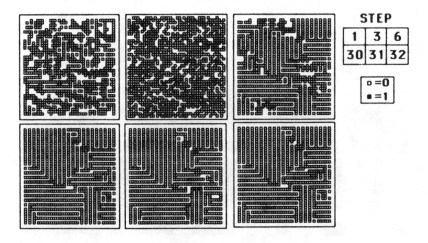

Figure 1.8. Synchronous iteration of the BNN for $|I| = 45$. The initial configuration converges to a cycle of period 2.

1.6.6. Bounded Majority Network. Let $\mathcal{A} = \left(I \times I, V_0^N, \{0,1\}, f\right)$ where

$$f\left(x_{(-1,0)}, x_{(0,1)}, x_{(0,0)}, x_{(1,0)}, x_{(0,-1)}\right) = \mathbb{1}\left(\sum_{(j,j') \in V_0^N} x_{(j,j')} - b\right)$$

As in the previous example, we suppose that all sites in $\mathbb{Z} \times \mathbb{Z} \setminus I \times I$ are fixed in state 0 (boundary condition).

Figure 1.9 shows the evolution for $|I| = 65$. The initial configuration evolves to rectangular patterns of $1'^s$ and two-cycles in a rectangular chess-board. This fact will be proved in Chapter 4.

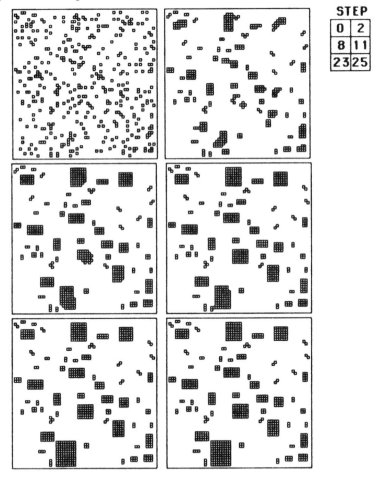

STEP	
0	2
8	11
23	25

Figure 1.9. Synchronous evolution of a Majority Network with the von Neumann neighbourhood, threshold $b = 2$, $|I| = 65$.

Now, for the same cellular space take the Moore neighbourhood, V_0^M, and the local function:

$$f(x_{(i,i')}; (i,i') \in V_0^M) = \mathbb{1}\left(\sum_{(j,j') \in V_0^M} x_{(j,j')} - b \right)$$

Figures 1.10 and 1.11 show the synchronous update for threshold 3 and 4 respectively. For $b = 3$ the $1'^s$ of the initial configuration quickly saturate the array with $1'^s$. For $b = 4$ initial patterns evolve to convex configurations of $1'^s$ which are stable.

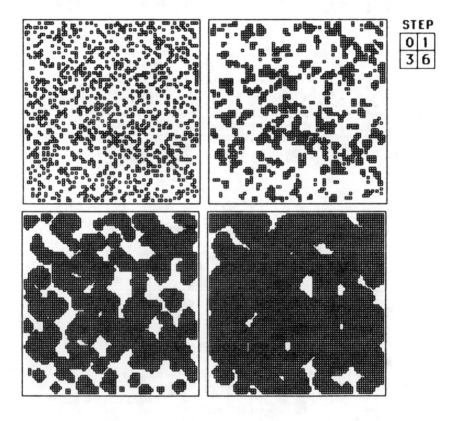

STEP

| 0 | 1 |
| 3 | 6 |

Figure 1.10. Synchronous evolution of a Majority Network with the Moore neighbourhood, $b = 3$, $|I| = 80$.

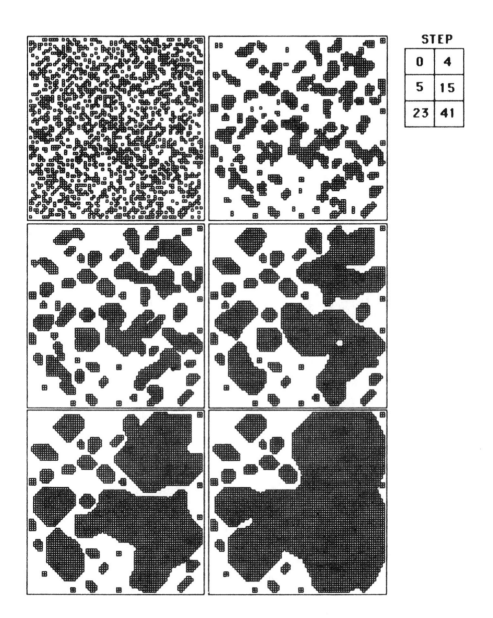

Figure 1.11. Synchronous evolution of a Majority Network with the Moore neighbourhood, $b = 4$, $|I| = 80$.

Majority Networks are a particular case of Neural Networks and they will be studied in more detail in the next chapter.

References

[AC] Albert, J., K. Culik, *A Simple Universal Cellular Automata and its one-way and Totalistic Version*, Complex Systems, 1(1), 1987, 1-16.

[BFW] Bienenstock, E., F. Fogelman-Soulie, G. Weisbuch, eds, *Disordered Systems and Biological Organization*, NATO ASI Series F: Comp. and System Sciences, Springer Verlag, 20, 1986.

[CY] Culik, K., S. Yu, *Undecidability of C.A. Classification Schemes*, Complex Systems, 2(2), 1988, 177-190.

[Ga2] Galperin, G.A., *One-Dimensional Monotonic Tesselations with Memory*, in Locally Interacting Systems and their Application in Biology, R.L. Dobrushin, V.I. Kryukov, A.L. Toom (eds), Lectures Notes in Mathematics N°653, Springer Verlag, 1978, 56-71.

[Gr] Green, F., *NP-Complete Problems in Cellular Automata*, Complex Systems, 1(3), 1987, 453-474.

[GT2] Goles, E., M. Tchuente, *Erasing Multithreshold Automata*, in Dynamical Systems and Cellular Automata, J. Demongeot, E. Goles, M. Tchuente (eds), Ac. Press, 1985, 47-56.

[He] Hedlund, G.A., *Endomorphism and Automorphisms of the Shift Dynamical System*, Math. System Theory, 3, 1969, 320-375.

[Ho2] Hopfield, J.J., *Neural Networks and Physical Systems with Emergent Collective Computational Abilities*, Proc. Nat. Acad. Sci., U.S.A., 79, 1982, 2554-2558.

[HoT] Hopfield, J.J., D.W. Tank, *Neural Computation of Decisions in Optimization Problems*, Biol. Cybern, 52, 1985, 141-152.

[Hu] Hurd, L., *Formal Language Characterisation of Cellular Automaton Limit Sets*, Complex Systems, 1(1), 1987, 69-80.

[K] Kleene, S.C., *Representation of Events in Nerve Nets and Finite Automata*, in Automata Studies, Annals of Mathematics Studies, 34, Princeton, 1956.

[L] Lind, D.A., *Applications of Ergodic Theory and Sofic Systems to Cellular Automata*, Physica, 10D, 1984, 36-44.

[LN] Lindgren, K., M. Nordahl, *Complexity Measures and Cellular Automata*, Complex Systems, 2(4), 409-440.

[Mi] Minsky, M.L., *Computation: Finite and Infinite Machines*, Prentice Hall, 1967.

[Ml] Milnor, J., *On the Entropy Geometry of Cellular Automata*, Complex Systems, 2(3), 1988, 257-385.

[MP] McCulloch, W.S., W. Pitts., *A Logical Calculus of the Ideas Immanent in Nervous Activity*, Bull. of Math. Biophys, 5, 1943, 115-133.

[Ri] Richardson, D., *Tessellation with Local Transformation*, J. Comput. & Systems Sci, 6, 1972, 373-388.

[RM] Rumelhart, D.E., J.C. McClelland (eds), *Parallel and Distributed Processing: Explorations in the Microstructure of Cognition*, MIT Press, 1986.

[Ro] Robert, F., *Discrete Iterations: A Metric Study*, Springer Series in Computational Mathematics, Springer-Verlag, 1986.

[Sm] Smith, A.R., *Simple Computation-Universal Cellular Spaces*, J.ACM, 18(3), 1971, 339-353.

[T1] Tchuente, M., *Contribution a l'Etude des Methodes de Calcul pour des Systemes de Type Cooperatif*, Thesis IMAG, Grenoble, France, 1982.

[ToM] Toom, A.L., L.G. Mitynshin, *Two Results Regarding Noncomputability for Univariate Cellular Automata*, Problemy peredaci Informacii, 12(2), 1976, 69-75.

[W1] Wolfram, S., *Theory and Applications of Cellular Automata*, World Scientific, 1986.

[W2] Wolfram, S., *Universality and Complexity in Cellular Automata*, Physica, 10D, 1984, 1-35.

2. ALGEBRAIC INVARIANTS ON NEURAL NETWORKS

2.1. Introduction

In this chapter we introduce mathematical tools, called algebraic invariants, which allow the characterization of the periodic behaviour of some classical models of neural computation. We include different ways of updating the networks: synchronous, sequential and block-sequential, which contains as particular cases the two previous ones. We also study memory iteration where the updating consider a longer history of each site. Finally, we also use algebraic invariantes to study majority networks.

To summarize the main idea of this chapter consider an evolution $F : C \to C$ on a finite set of configurations C. A trajectory of such a system is of the form:

$$x(t+1) = F(x(t)) \quad t \geq 0, \quad x(t) \in C$$

From the finiteness of C any trajectory is ultimately periodic, i.e. converges (in a finite number of steps) to a cycle $X = (x(0), ..., x(T-1))$, where T is the period of the system. We define a quantity $L(X)$ -called an algebraic invariant- driving the dynamics in the steady state X, and whose properties enable the characterization of the period of relevant classes of Neural Networks. In the general case the quantity $L(X)$ gives only partial information about the period of the system.

Although we could present algebraic invariants in a more compact form, we have prefered to develop some relevant cases and to postpone a general framework to next chapters. This is also the case for transient properties which will be studied in Chapter 3, once powerful tools have been introduced.

2.2. k-Chains in 0-1 Periodic Sequences

In this paragraph we shall introduce some concepts which will be useful when we relate the algebraic invariants to the periods of the limit orbits.

Let $Y = (y(t) : t \in \mathbb{N})$ be a periodical sequence of 0's and 1's; suppose its period $\gamma(Y)$ (which is a priori unknown) divides T. Thus $y(t) \in \{0,1\}$ for any $t \in \mathbb{Z}$ and $y(t) = y(t')$ when $t \equiv t' \pmod{T}$. Obviously, we can extend the domain of definition of Y to $\mathbb{Z}$; furthermore, Y is well defined over $\mathbb{Z}_T = \mathbb{Z}/\text{mod } T$ (the cyclic group of integers mod T). Then, for all purposes concerning the study of the period of the sequence, we assume that t takes its values on $\mathbb{Z}_T$, and the sequence will be defined in this set, $Y = (y(t) : t \in \mathbb{Z}_T)$.

In studying period lengths we shall deal with sets invariant under translations, so the following notation will be useful: if $\Gamma \subset \mathbb{Z}_T$, $l \in \mathbb{Z}$ we write:

$$\Gamma + l = \{t + l(\text{ mod } T) : t \in \Gamma\}.$$

Let us partition the set $\mathbb{Z}_T$ into $\Gamma^0(Y) = \{t \in \mathbb{Z}_T : y(t) = 0\}$ and $\Gamma^1(Y) = \{t \in \mathbb{Z}_T : y(t) = 1\}$, which is called the support of Y. The period of the set $\Gamma^1(Y)$ is the smallest positive number γ such that $\Gamma^1(Y) + \gamma = \Gamma^1(Y)$. A first obvious remark is that $\gamma(Y)$, the period of the sequence, is equal to the period of $\Gamma^1(Y)$. Then:

$$\gamma(Y)|k \quad \text{iff} \quad \Gamma^1(Y) + k = \Gamma^1(Y) \tag{2.1}$$

where $\gamma'|\gamma''$ means γ' divides γ''.

Now let us define k-chains (for $k \geq 1$) contained in the support $\Gamma^1(Y)$. A subset $C \subset \Gamma^1(Y)$ is called a k-*chain* iff it is of the form $C = \{t + kl(\bmod T) : 0 \leq l < s\}$ for some $s \geq 1$. So a k-chain is a subset $C = \{t + kl \in \mathbb{Z}_T : 0 \leq l < s\}$ such that $y(t') = 1$ for any $t' \in C$.

A k-chain is said to be *maximal* if it is not strictly contained in another k-chain. Obviously, everyone of the k-chains in $\Gamma^1(Y)$ is contained in a maximal k-chain. Let $\varsigma(Y)$ be the class of maximal k-chains; from the definitions, it partitions the support $\Gamma^1(Y)$, i.e.

$$\bigcup_{C \in \varsigma(Y)} C = \Gamma^1(Y) \quad \text{and} \quad C \cap C' = \phi$$

for any couple $C \neq C'$ of different maximal k-chains in $\varsigma(Y)$.

Maximal k-chains differ qualitatively, according as they are invariant under translations of order k or not, that is, whether they satisfy $C = C + k$ or not. Recall that $C = C + k$ is equivalent to $C = C + kl$ for any $l \in \mathbb{Z}$, as is equivalent to $C = C - k$.

Let $\varsigma^{(k)}(Y) = \{C \in \varsigma(Y) : C = C + k\}$ be the class of k-periodic k-chains. Then:

Lemma 2.1. $\gamma(Y)|k$ iff $\varsigma^{(k)}(Y) = \varsigma(Y)$ (this means that every maximal k-chain is k-periodic).

Proof. Assume $\varsigma^{(k)}(Y) = \varsigma(Y)$, let $t \in \Gamma^1(Y)$. If C is the maximal k-chain containing t we have $C = C + k$, so $t + k \pmod T \in C \subset \Gamma^1(Y)$. Then $\Gamma^1(Y) = \Gamma^1(Y) + k$. From (2.1) we deduce $\gamma(Y)|k$.

Suppose $\exists C \in \varsigma(Y) \setminus \varsigma^{(k)}(Y)$, then there exists $t \in C \setminus C - k$, i.e. $t \in C$ and $t + k(\bmod T) \notin C$. By maximality of C this implies $t + k(\bmod T) \notin \Gamma^1(Y)$. Then $\Gamma^1(Y) + k \neq \Gamma^1(Y)$, which is equivalent to the assertion that $\gamma(Y)$ does not divide k. ∎

Note that $\gamma(Y) = 1$ iff $\Gamma^1(Y) = \mathbb{Z}_T$ or ϕ (then $\varsigma(Y) = \varsigma^{(1)}(Y) = \{\mathbb{Z}_T\}$ or ϕ respectively), and $\gamma(Y) = 2$ iff there exists a unique maximal 2-chain different from $\mathbb{Z}_T$ and ϕ.

Now call s_c the cardinality of $C \in \varsigma(Y)$. If $C \in \varsigma^{(k)}(Y)$ choose anyone of its elements and call it $\underline{t}_c$. Otherwise, i.e. $C \in \varsigma(Y) \setminus \varsigma^{(k)}(Y)$, take $\underline{t}_c$ such that $\underline{t}_c - k(\bmod T) \notin C$ (being C a k-chain $\underline{t}_c$ is necessarily unique). Then

$$C = \{\underline{t}_c + lk(\bmod T) : 0 \le l < s_c\}.$$

For any $C \in \varsigma(Y)$ write $\bar{t}_c = \underline{t}_c + (s_c - 1)k(\bmod T)$ which is an element of C.
From definitions (in particular employing the maximality) we have:

$$\begin{cases} C \in \varsigma^{(k)}(Y) \text{ iff } \bar{t}_c + k(\bmod T) \in \Gamma^1(Y) \text{ iff } \underline{t}_c - k(\bmod T) \in \Gamma^1(Y) \\ \text{iff } \bar{t}_c + k(\bmod T) = \underline{t}_c \text{ iff } \underline{t}_c - k(\bmod T) = \bar{t}_c \end{cases} \quad (2.2)$$

So when $k = 2$ we have $C \in \varsigma^{(2)}(Y)$ iff $\underline{t}_c - 1 \equiv \bar{t}_c + 1(\bmod T)$

To illustrate above concepts we shall exhibit two examples.
Consider the periodic sequence $Y = (0,0,1,1,0,0,1,1,0,0,1,1)$ with $T = 12$. So $\Gamma^0(Y) = \{0,1,4,5,8,9\}$, $\Gamma^1(Y) = \{2,3,6,7,10,11\}$.
Take $k = 4$. The set of maximal 4-chains contained in $\Gamma^1(Y)$ is $\varsigma = \{C, C'\}$ where $C = \{2,6,10\}$, $C' = \{3,7,11\}$. Both chains C, C' are 4-periodic so $\varsigma^{(4)} = \varsigma$ then the period $\gamma(Y)$ of Y divides $k = 4$. Obviously $\gamma(Y) = 4$ in this case. The cardinality of both chains is $s_c = s_{c'} = 3$. If we choose $\underline{t}_c = 2$, $\underline{t}_{c'} = 7$ we find $\bar{t}_c = 10$, $\bar{t}_{c'} = 3$.
For $k = 5$ the set of maximal 5-chains in $\Gamma^1(Y)$ is $\varsigma(Y) = \{C, C', C''\}$ where $C = \{2,7\}$, $C' = \{10,3\}$, $C'' = \{6,11\}$. None of these chains is 5-periodic. For instance for C we have $\underline{t}_c = 10$, $\underline{t}_c - 5(\bmod 12) = 5$ which does not belong to $\Gamma^1(Y)$. Note that if a maximal 5-chain existed which is 5-periodic then necessarily $\Gamma^1(Y) = \mathbb{Z}_{12}$ because 5 and 12 are relatively prime.
Now consider the sequence $Y = (0,1,0,1,1,1,1,1,1,1,0,1)$ whose support is $\Gamma^1(Y) = \{1,3,4,5,6,7,8,9,11\}$. Take $k = 2$. We have $\varsigma(Y) = \{C, C'\}$ where $C = \{1,3,5,7,9,11\}$ and $C' = \{4,6,8\}$. Note that $\varsigma^{(2)}(Y) = \{C\}$, $\varsigma(Y) \setminus \varsigma^{(2)}(Y) = \{C'\}$, so for a sequence Y which is not 2-periodic we may have $\varsigma^{(2)}(Y) \neq \phi$.
For C, its cardinality is $s_c = 6$. We can choose $\underline{t}_c = 1$, then $\bar{t}_c = 11$. Since C is a maximal 2-periodic chain $\underline{t}_c - 1(\bmod 12) = 11 \in \Gamma^1(Y)$. For C' we have $s_{c'} = 3$, $\underline{t}_{c'} = 4$ and $\bar{t}_{c'} = 8$. In this case $\underline{t}_{c'} - 2 = 2 \notin \Gamma^1(Y)$.

2.3. Covariance in Time

One of the most important parameters for understanding a Neural Network dynamics is the covariance between the processes in different sites, as a time function. More precisely, consider a Neural Network whose states evolve according to an equation $x(t+1) = F(x(t))$ for $t \geq 0$ or $x(t+1) = \bar{F}(x(t), ..., x(t-(r-1)))$ for $t \geq r - 1 \geq 0$.

Note that this last class of evolutions can be conveniently coded as $y(t+1) = F(y(t))$ for certain F.

If the neuron at site $i \in I = \{1, ..., n\}$ is excited at time t we put $x_i(t) = 1$, and if it is inhibited, $x_i(t) = 0$. Then the states of the network $x(t) = (x_i(t) : i \in I)$ take values on $\{0, 1\}^n$. We shall be interested in studying the joint processes $(x_i(t)x_j(t+k) : t \geq 0)$ by functions which measure the predictibility, at k steps of time, of the neuron value at site j by knowing the neuron state at site i. The functions considered will be the average of the product process in long intervals of time. As the state space $\{0, 1\}^n$ is finite, any initial condition converges in a finite number of steps to finite cycles. Then it suffices for us to study the covariance among $(x_i(t), x_j(t-k))$ when we have reached the limit regime.

Let $x(0)$ be the initial condition of the system. The covariance of the limit process in sites i, j at k steps of time is:

$$V^k(x(0), i, j) = \lim_{t \to \infty} \left\{ \left(\frac{1}{t} \sum_{s=0}^{t-1} x_i(s)x_j(s+k) \right) - \left(\frac{1}{t} \sum_{s=0}^{t-1} x_i(s) \right) \left(\frac{1}{t} \sum_{s=0}^{t-1} x_j(s+k) \right) \right\}$$

(2.3)

The above quantity is defined for any $k \in \mathbb{Z}$, if $k < 0$ we begin the sum at $s = |k|$. Suppose $x(t)$ enters at a certain t_0 a cycle of length T: $(x(t_0), ..., x(t_0 + T - 1))$. Obviously, the time covariance function can be described by using the values of the processes in the cycle. Shift the time in t_0, that is take $x(0) = x(t_0)$, then the evolution $x(t)$ is well defined in $\mathbb{Z}_T = \mathbb{Z}/\mod T$. The T-cycle at site i is denoted $X_i = (x_i(t) : t \in \mathbb{Z}_T)$, and the mean limit activity of neuron at site i is $\bar{X}_i = \frac{1}{T} \sum_{t \in \mathbb{Z}_T} x_i(t)$. Then the above quantity $V^k(x(0), i, j)$ can also be written as

$$V^k(X_i, X_j) = \frac{1}{T} \sum_{t \in \mathbb{Z}_T} x_i(t)x_j(t+k) - \bar{X}_i \bar{X}_j$$

(2.4)

and by definition it is T-periodic in k: $V^{k+T}(\cdot, \cdot) = V^k(\cdot, \cdot)$ and,

$$V^k(X_i, X_j) = V^{-k}(X_j, X_i) = V^{T-k}(X_j, X_i)$$

(2.5)

For the purpose of studying the periodic behavior of a large class of Neural Networks, the differences between these covariances in time are relevant. Define:

$$\Delta V^{k,l}(X_i, X_j) = V^k(X_i, X_j) - V^l(X_j, X_i) \text{ for } k, l \in \mathbb{Z}$$

(2.6)

which is T-periodic in k, l : $\Delta V^{k+T,l}(\cdot, \cdot) = \Delta V^{k,l+T}(\cdot, \cdot) = \Delta V^{k,l}(\cdot, \cdot)$ and

$$\Delta V^{k,l}(X_i, X_j) = -\Delta V^{l,k}(X_j, X_i) \tag{2.7}$$

Now denote by $\gamma(X_i)$ the period of the i-th cycle X_i, by definition $\gamma(X_i)|T$. An important relation among the difference of covariances and the period is given by:

Lemma 2.2. If $\gamma(X_i)|k + l$ then $\Delta V^{k,l}(X_i, X_j) = 0$ for any $j \in I$.

Proof. Let $\Gamma^0(X_i) = \{t \in \mathbb{Z}_T : x_i(t) = 0\}$, $\Gamma^1(X_i) = \{t \in \mathbb{Z}_T : x_i(t) = 1\}$, $\varsigma(X_i)$ be the class of maximal $k + l$-chains contained in $\Gamma^1(X_i)$, and $\varsigma^{(k+l)}(X_i)$ be the subclass of $k + l$-periodic maximal chains.
We have:

$$\Delta V^{k,l}(X_i, X_j) = \frac{1}{T} \sum_{t \in \mathbb{Z}_T} x_i(t)(x_j(t + k) - x_j(t - l)) \tag{2.8}$$

so $\Delta V^{k,l}(X_i, X_j) = \frac{1}{T} \sum_{C \in \varsigma(X_i)} (\sum_{t \in C} x_j(t + k) - x_j(t - l))$. Using notation from previous sections, we write $C = \{\underline{t}_c + r(k + l)(\bmod T) : 0 \le r < s_c\}$ where s_c is the cardinality of C and $\bar{t}_c = \underline{t}_c + (s_c - 1)(k + l)$. Then:

$$\Delta V^{k,l}(X_i, X_j) = \frac{1}{T} \sum_{C \in \varsigma(X_i)} (x_j(\bar{t}_c + k) - x_j(\underline{t}_c - l)) \tag{2.9}$$

For any $C \in \varsigma^{(k+l)}(X_i)$ we have $\bar{t}_c + k \equiv \underline{t}_c - l(\bmod T)$ so:

$$\Delta V^{k,l}(X_i, X_j) = \frac{1}{T} \sum_{C \in \varsigma(X_i) \setminus \varsigma^{(k+l)}(X_i)} (x_j(\bar{t}_c + k) - x_j(\underline{t}_c - l)) \tag{2.10}$$

A sum over a void set of indexes is null, so $\varsigma(X_i) = \varsigma^{(k+l)}(X_i)$ implies that the above quantity is null, i.e. $\Delta V^{k,l}(X_i, X_j) = 0$.
As $\gamma(X_i)|k + l$ implies $\varsigma(X_i) = \varsigma^{(k+l)}(X_i)$ (see lemma 2.1) we deduce $\Delta V^{k,l}(X_i, X_j) = 0$ ∎

2.4. Algebraic Invariants of Synchronous Iteration on Neural Networks

Let $I = \{1, ..., n\}$ be the set of sites i of the Neural Network. We denote by $x(t) = (x_i(t) : i \in I) \in \{0, 1\}^n$ the state of the network at each time $t \geq 0$, where

$$x_i(t) = \begin{cases} 1 & \text{if the neuron at site } i \text{ is excited at time } t, \\ 0 & \text{if the neuron at site } i \text{ is inhibited at time } t, \end{cases}$$

Let a_{ij} be the interaction between the neurons at sites i and j, and b_i be the threshold of excitation of the neuron at site i.

The synchronous iteration of such a Neural Network is given by the equation:

$$x_i(t+1) = \mathbb{1}\left(\sum_{j \in I} a_{ij} x_j(t) - b_i\right), \quad i \in I, \ t \geq 0 \tag{2.11}$$

where $\mathbb{1}(u)$ is the threshold function introduced in (1.5).

Let $A = (a_{ij} : i, j \in I)$ be the matrix of interactions, $b = (b_i : i \in I)$ the vector of thresholds and $\bar{\mathbb{1}}$ the multidimensional threshold function. The evolution equation (2.11) of Neural Network $N = (I, a, b)$ can be written in the same form as in (1.7):

$$x(t+1) = F_N(x(t)) \quad \text{with } F_N(x) = \bar{\mathbb{1}}(Ax - b) \tag{2.12}$$

Since the network evolves in a finite set, $\{0, 1\}^n$, the synchronous iteration defined by equation (2.11) converges, for any initial configuration, in a finite number of steps to a steady-state, that is to a finite cycle. Let us call T the period of the Neural Network, that is the smallest number that is exactly divisible by the period of each finite cycle of iteration (2.11).

Let $X = (x(0), ..., x(T-1))$ be a finite T-cycle, that is $x(t) = x(t')$ if $t \equiv t' (\bmod T)$. Then $x(t)$ is defined for $t \in \mathbb{Z}_T$ and satisfies

$$x(t+1) = \bar{\mathbb{1}}(Ax(t) - b) \quad \text{for any } t \in \mathbb{Z}_T.$$

We denote by X_i the T-cycle described by site i: $X_i = (x_i(t) : t \in \mathbb{Z}_T)$, and by $\gamma(X_i)$ its period, which, by definition, divides T.

The synchronous invariant defined in limit cycles will be the difference of covariances at one-step, weighted by the interaction. When the interactions are symmetric it will be null in the limit cycle. The synchronous functional between the cycles X_i, X_j is:

$$L_{sy}(X_i, X_j) = a_{ij} \Delta V^{1,1}(X_i, X_j) \tag{2.13}$$

which can be written as (see (2.8)):

$$L_{sy}(X_i, X_j) = a_{ij} \frac{1}{T} \sum_{t \in \mathbb{Z}_T} x_i(t)(x_j(t+1) - x_j(t-1)) \qquad (2.14)$$

From results of section 2.3 we can deduce some properties which do not depend on the particular evolution (2.11) on $\{0,1\}^n$, i.e. which are true for any evolution $x(t+1) = F(x(t))$ (or $x(t+1) = \bar{F}(x(t), ..., x(t-(r-1))))$. From (2.7) we get:

$$\text{if } a_{ij} = a_{ji} \text{ then } L_{sy}(X_i, X_j) + L_{sy}(X_j, X_i) = 0 \qquad (2.15)$$

and from lemma 2.2 we obtain the first relationship between the period and L_{sy}:

$$\text{if } \quad \gamma(X_i)|2 \quad \text{then} \quad L_{sy}(X_i, X_j) = 0 \quad \forall j \in I \qquad (2.16)$$

From the particular form of the evolution equation (2.11) we can obtain finer results:

Lemma 2.3. The synchronous functional satisfies the following properties:

for any interaction matrix A we have: $\sum_{j \in I} L_{sy}(X_i, X_j) \le 0 \quad \forall i \in I \qquad (2.17)$

and the relationships between the functional and the periods are given by:

$$L_{sy}(X_i, X_j) = 0 \text{ for any } j \in I \text{ iff } \sum_{j \in I} L_{sy}(X_i, X_j) = 0 \text{ iff } \gamma(X_i)|2 \qquad (2.18)$$

$$\sum_{j \in I} L_{sy}(X_i, X_j) < 0 \text{ iff } \gamma(X_i) > 2 \qquad (2.19)$$

$$\sum_{i \in I} \sum_{j \in I} L_{sy}(X_i, X_j) = 0 \text{ iff } \gamma(X_i)|2 \text{ for any } i \in I \qquad (2.20)$$

Proof. Note that property (2.20) follows from (2.17) - (2.19). As in section 2.2, partition $\mathbb{Z}_T$ into the sets $\Gamma^0(X_i) = \{t \in \mathbb{Z}_T : x_i(t) = 0\}$ and the support of the cycle $\Gamma^1(X_i) = \{t \in \mathbb{Z}_T : x_i(t) = 1\}$. Let $\varsigma(X_i)$ be the class of maximal 2-chains contained in the support and $\varsigma^{(2)}(X_i) = \{C \in \varsigma(X_i) : C - 2 = C\}$ be the class of 2-periodic maximal 2-chains. With notations of section 2.2, a maximal 2-chain $C \in \varsigma(X_i)$ is written $C = \{\underline{t_c} + 2l(\bmod T) : 0 \le l < s_c\}$ where s_c is the cardinality of C, $\underline{t_c}$ is anyone of the elements of C when $C \in \varsigma^{(2)}(X_i)$ or the unique element of C such that $\underline{t}_c - 2(\bmod T) \notin C$ when $C \in \varsigma(X_i) \setminus \varsigma^{(2)}(X_i)$. We also note $\bar{t}_c = \underline{t}_c + 2(s_c - 1)(\bmod T)$.

From (2.10) we have $L_{sy}(X_i, X_j) = \frac{a_{ij}}{T} \sum\limits_{C \in \varsigma(X_i) \setminus \varsigma^{(2)}(X_i)} (x_j(\bar{t}_c + 1) - x_j(\underline{t}_c - 1))$,

hence

$$\sum_{j \in I} L_{sy}(X_i, X_j) = \frac{1}{T} \sum_{C \in \varsigma(X_i) \setminus \varsigma^{(2)}(X_i)} \left\{ \sum_{j \in I} a_{ij} x_j(\bar{t}_c + 1) - \sum_{j \in I} a_{ij} x_j(\underline{t}_c - 1) \right\}$$

$$(2.21)$$

As $\underline{t}_c \in \Gamma^1(X_i)$ we have $x_i(\underline{t}_c) = 1$. By the special form of our dynamical equation (2.11), this is equivalent to $\sum\limits_{j \in I} a_{ij} x_j(\underline{t}_c - 1) \geq b_i$.

When $C \in \varsigma(X_i) \setminus \varsigma^{(2)}(X_i)$ we necessarily have (see (2.2)) $\bar{t}_c + 2(\bmod T) \notin \Gamma^1(X_i)$, so $x_i(\bar{t}_c + 2) = 0$. Then we must have $\sum\limits_{j \in I} a_{ij} x_j(\bar{t}_c + 1) < b_i$. Hence:

$$\text{for any } C \in \varsigma(X_i) \setminus \varsigma^{(2)}(X_i) : \sum_{j \in I} a_{ij} x_j(\bar{t}_c + 1) - \sum_{j \in I} a_{ij} x_j(\underline{t}_c - 1) < 0 \quad (2.22)$$

From (2.21), (2.22) we deduce $\sum\limits_{j \in I} L_{sy}(X_i, X_j) \leq 0$, $L_{sy}(X_i, X_j) = 0$ for any $j \in I$ iff $\varsigma(X_i) = \varsigma^{(2)}(X_i)$ and $\sum\limits_{j \in I} L_{sy}(X_i, X_j) < 0$ iff $\varsigma(X_i) \setminus \varsigma^{(2)}(X_i) \neq \phi$.

The assertions of lemma 2.1 imply our result. ∎

Hence (2.20) of lemma 2.3 expresses that $\sum\limits_{i \in I} \sum\limits_{j \in I} L_{sy}(X_i, X_j) = 0$ for any cycle X is a necessary and sufficient condition for $T|2$. From (2.15), the condition that A is symmetric implies $\sum\limits_{i \in I} \sum\limits_{j \in I} L_{sy}(X_i, X_j) = 0$ for each cycle X, so:

Theorem 2.1. [GO3] If the matrix of interactions A is symmetric then the period T of the network $\mathcal{N} = (I, A, b)$ iterated synchronously satisfies $T|2$. ∎

Remarks.
1. When A is not symmetric we can get limit cycles of long periods. Let us show some examples where cycles of non-bounded periods are obtained (i.e. the period $T_n \to \infty$ when $n \to \infty$, n being the size of the network).

Take the following matrix of interactions,

$$A = \begin{pmatrix} 0 & 0 & \dots & 0 & 1 \\ 1 & 0 & \dots & 0 & 0 \\ 0 & 1 & \dots & 0 & 0 \\ 0 & 0 & \dots & 1 & 0 \end{pmatrix}$$

and the constant threshold $b_i = \frac{1}{2}$ for any $i \in I$. Then the evolution is $x_i(t+1) = \mathbb{1}\left(x_{i-1}(t) - \frac{1}{2}\right)$ for $i = 1, ..., n$ where the indexes are mod n i.e. $0 \equiv n$. Denote $e_r = (0, ...1, ...0)$, where the '1' is at the r-th place. It is easy to see that for any $r = 1, ..., n$ the initial condition $x(0) = e_r$ belongs to a cycle of length n, in fact it evolves as follows: $e_r \rightarrow e_{r+1} \rightarrow ... \rightarrow e_{r-1} \rightarrow e_r$.

2. Furthermore, we can construct examples having non-bounded periods for inter-action matrices which are arbitrarily close to symmetric matrices. More precisely, cycles of non-bounded periods will be constructed for all the matrices belonging to a one-parameter class $(A(\epsilon) : \epsilon > 0)$. Each one of these matrices $A(\epsilon)$ will satisfy: $a_{ij}(\epsilon) = a_{ji}(\epsilon)$ for any pair $\{i, j\} \neq \{i^*, j^*\}$, while for the unique pair $\{i^*, j^*\}$, where symmetry is not satisfied, we shall have $0 < |a_{i^*j^*}(\epsilon) - a_{j^*i^*}(\epsilon)| \xrightarrow{\epsilon \to 0} 0$.

Let us construct $A(\epsilon) = (a_{ij}(\epsilon) : i, j = 1, ..., n)$. The set of indexes $\{\overset{\epsilon \to 0}{1}, ..., n\}$ is defined mod n, so $0 \equiv n$, $1 \equiv n+1$. Take:

$$a_{ij}(\epsilon) = \begin{cases} 0 & \text{if } i-1 \neq j \neq i+1, \\ 1 - i\epsilon & \text{if } j = i+1 \\ 1 - (i-1)\epsilon & \text{if } j = i-1 \end{cases} \qquad \text{when } \{i, j\} \neq \{1, n\},$$

and $a_{1,n}(\epsilon) = 1$, $a_{n,1}(\epsilon) = 1 - n\epsilon$.

Note that $a_{ij}(\epsilon) = a_{ji}(\epsilon)$ for each $\{i, j\} \neq \{1, n\}$ and $|a_{1,n}(\epsilon) - a_{n,1}(\epsilon)| \xrightarrow{\epsilon \to 0} 0$. The threshold values are $b_i = 1 - (i-1)\epsilon$ for $i = 1, ..., n$. It is easy to see that the evolution $x(t+1) = \bar{\mathbb{1}}(Ax(t) - b)$ is such that $e_1 \rightarrow e_2 \rightarrow ... \rightarrow e_n \rightarrow e_1$, hence there exist cycles of length n.

From this example we can see that there exists a very strong tradeoff among entries a_{ij} and thresholds b_i, in fact threshold values introduce asymmetries in the network.

3. It is not difficult to see that for the following synchronous iteration:

$$x(t+1) = \bar{\mathbb{1}}(Ax(t) - b_1) \text{ for } t \text{ even,}$$
$$x(t+1) = \bar{\mathbb{1}}(Ax(t) - b_2) \text{ for } t \text{ odd}$$

we also have the two-periodic behaviour of limit cycles when A is symmetric. To show the assertion it suffices to follow the parity of the maximal 2-chain partition.

4. We may generalize Theorem 2.1 to matrices A satisfying:

there exists a permutation matrix P such that $PA\,{}^tP = \begin{bmatrix} A_1 & & \\ 0 & \ddots & \\ 0 & 0 & A_s \end{bmatrix}$

where the square matrices $A_1, ..., A_s$ are symmetric. This generalization is proved by using Theorem 2.1 and the previous remark.

2.5. Algebraic Invariants of Sequential Iteration on Neural Networks

In a sequential iteration the set of sites I is totally ordered. We shall assume, without loss of generality, that the order on I is the same order as I posseses as a subset of $\mathbb{Z}$.

Besides, the update of the neurons when the network evolves from t to $t + 1$ occurs hierarchically. Thus, when the neuron i changes from $x_i(t)$ to $x_i(t + 1)$, all the neurons at sites $j < i$ have already evolved. The meaning of this evolution is analogous to that of the synchronous one studied in section 2.4, that is $x_i(t+1) = 1$ is active iff the interactions (which are assumed linear, as given by a matrix A) of the active neurons measured at i exceed its threshold. The neurons considered for the interactions are in this case those active at $t + 1$ on sites $j < i$, and the neurons active at t on sites $j \geq i$.

Thus the configurations of our system are $x(t) \in \{0, 1\}^n$, the matrix of interactions is $A = (a_{ij} : i, j \in I = \{1, ..., n\})$ and the threshold vector is $b = (b_i : i \in I)$. The sequential updating of the Neural Network is written:

$$x_i(t + 1) = \mathbb{1}\left(\sum_{j=1}^{i-1} a_{ij} x_j(t + 1) + \sum_{j=i}^{n} a_{ij} x_j(t) - b_i\right) \qquad (2.23)$$

(since a sum over an empty set of indexes is null, $\sum_{j=1}^{i-1} = 0$ if $i = 1$).

Let T be the period of the Neural Network. Let $X = (x(0), ..., x(T - 1))$ be a T-cycle; then $(x(t) : t \in \mathbb{Z}_T)$ is well defined in $\mathbb{Z}_T : x(t) = x(t')$ if $t \equiv t'(\mod T)$, and $x(t + 1)$ obeys equations (2.24) for any $t \in \mathbb{Z}_T$.

We denote by $X_i = (x_i(t) : t \in \mathbb{Z}_T)$ the i-th cycle and by $\gamma(X_i)$ its period, which is a divisor of T.

The sequential functional will measure the difference of covariances between consecutive and equal time steps, weighted by the interaction. So for any couple of local cycles (X_i, X_j) we define:

$$L_{se}(X_i, X_j) = \begin{cases} a_{ij} \Delta V^{1,0}(X_i, X_j) & \text{if } j < i \\ a_{ij} \Delta V^{0,0}(X_i, X_j) & \text{if } j = i \\ a_{ij} \Delta V^{0,1}(X_i, X_j) & \text{if } j > i \end{cases} \qquad (2.24)$$

If we develop this quantity we find:

$$L_{se}(X_i, X_j) = \begin{cases} \frac{a_{ii}}{T} \sum_{t \in \mathbb{Z}_T} x_i(t)(x_j(t + 1) - x_j(t)) & \text{if } j > i \\ 0 & \text{if } j = i \\ \frac{a_{ii}}{T} \sum_{t \in \mathbb{Z}_T} x_i(t)(x_j(t) - (x_j(t - 1)) & \text{if } j > i \end{cases} \qquad (2.25)$$

As with L_{sy}, the results of section 2.3 lead to results which are true not only for evolution (2.23) but for any dynamical equation: $x(t+1) = F(x(t))$ (or $x(t+1) = \bar{F}(x(t), ..., x(t-(r-1))))$). Thus (2.7) implies:

$$\text{if } a_{ij} = a_{ji} \text{ then } L_{se}(X_i, X_j) + L_{se}(X_j, X_i) = 0 \qquad (2.26)$$

and from lemma 2.2:

$$\text{if } \gamma(X_i) = 1 \text{ then } L_{se}(X_i, X_j) = 0 \text{ for any } j \in I. \qquad (2.27)$$

For the evolution (2.23) we obtain the following relationship between the period $\gamma(X_i)$ and the invariant L_{se}:

Lemma 2.4. Let A be an interaction matrix such that $a_{ii} \geq 0$ for any $i \in I$. Then:

$$\sum_{j \in I} L_{se}(X_i, X_j) \leq 0 \qquad (2.28)$$

$$L_{se}(X_i, X_j) = 0 \text{ for any } j \in I \text{ iff } \sum_{j \in I} L_{se}(X_i, X_j) = 0 \text{ iff } \gamma(X_i) = 1 \qquad (2.29)$$

$$\sum_{j \in I} L_{se}(X_i, X_j) < 0 \text{ iff } \gamma(X_i) > 1 \qquad (2.30)$$

$$\sum_{i \in I} \sum_{j \in I} L_{se}(X_i, X_j) = 0 \text{ iff } \gamma(X_i) = 1 \text{ for any } i \in I \qquad (2.31)$$

Proof. Property (2.31) is deduced from (2.28) - (2.30). Partition $\mathbb{Z}_T$ into $\Gamma^0(X_i) = \{t \in \mathbb{Z}_T : x_i(t) = 0\}$ and $\Gamma^1(X_i) = \{t \in \mathbb{Z} : x_i(t) = 1\}$. Denote by $\varsigma(X_i)$ the set of maximal 1-chains and by $\varsigma^{(1)}(X_i)$ the set of 1-periodic maximal 1-chains. Obviously if $\varsigma^{(1)}(X_i) \neq \phi$ we have $\varsigma^{(1)}(X_i) = \{\mathbb{Z}_T\}$. In this case $\gamma(X_i) = 1$. Analogously, if $\Gamma^1(X_i) = \phi$ we also have $\gamma(X_i) = 1$. These two are the only cases where $\gamma(X_i) = 1$.

Then assume $\gamma(X_i) > 1$ so any maximal 1-chain $C \in \varsigma(X_i)$ is not 1-periodic $(C \neq C + 1)$ and $\varsigma^{(1)}(X_i) = \phi$. Following notation of section 2.2 we write $C = \{\underline{t}_c + l(\mod T) : 0 \leq l < s_c\}$ and $\bar{t}_c = \underline{t}_c + s_c - 1(\mod T)$. As $C \neq C + 1$ we have $\bar{t}_c + 1(\mod T) \notin C$.

From (2.9) we have:

$$T \sum_{j \in I} L_{se}(X_i, X_j) =$$

$$\sum_{C \in \varsigma(X_i)} \left(\sum_{j=1}^{i-1} a_{ij}(x_j(\bar{t}_c + 1) - x_j(\underline{t}_c)) + \sum_{j=i+1}^{n} a_{ij}(x_j(\bar{t}_c) - x_j(\underline{t}_c - 1)) \right) =$$

$$\sum_{C \in \varsigma(X_i)} \left(\left(\sum_{j=1}^{i-1} a_{ij} x_j(\bar{t}_c + 1) + \sum_{j=i}^{n} a_{ij} x_j(\bar{t}_c) \right) - \left(\sum_{j=1}^{i-1} a_{ij} x_j(\underline{t}_c) + \sum_{j=i}^{n} a_{ij} x_j(\underline{t}_c - 1) \right) \right)$$

$$- \sum_{C \in \varsigma(X_i)} a_{ii}(x_i(\bar{t}_c) - x_i(\underline{t}_c - 1))$$

Now $x_i(\underline{t}_c) = 1$, so the form of the iteration (2.23) implies

$$\sum_{j=1}^{i-1} a_{ij} x_j(\underline{t}_c) + \sum_{j=i}^{n} a_{ij} x_j(\underline{t}_c - 1) \geq b_i.$$

On the other hand, each $C \in \varsigma(x_i)$ is not 1-periodic so $x_i(\bar{t}_c + 1) = 0$ which means $\sum_{j=1}^{i-1} a_{ij} x_j(\bar{t}_c + 1) + \sum_{j=i}^{n} a_{ij} x_j(\bar{t}_c) < b_i$. Then

$$\sum_{j \in I} L_{se}(X_i, X_j) < -\frac{a_{ii}}{T} \sum_{C \in \varsigma(X_i)} (x_i(\bar{t}_c) - x_i(\underline{t}_c - 1)).$$

But $x_i(\bar{t}_c) = 1$, $x_i(\underline{t}_c - 1) = 0$ so $\sum_{C \in \varsigma(X_i)} (x_i(\bar{t}_c) - x_i(\underline{t}_c - 1)) > 0$. Hence condition $a_{ii} \geq 0$ implies $\sum_{j \in I} L_{se}(X_i, X_j) < 0.$ ∎

Suppose A has non-negative diagonal entries. From (2.31) of lemma 2.4 we find that the condition $\sum_{i \in I} \sum_{j \in I} L_{se}(X_i, X_j) = 0$ for any cycle X is necessary and sufficient to get $T = 1$. Besides, (2.26) implies that A symmetric is a sufficient condition to have $\sum_{i \in I} \sum_{j \in I} L_{se}(X_i, X_j) = 0$ for any cycle X, so:

Theorem 2.2. [G5,G7] Assume the matrix of interactions A to be symmetric with non-negative diagonal entries $a_{ii} \geq 0$ for any $i \in I$. Then the period T of the sequential iteration of the Neural Network $\mathcal{N} = (I, A, b)$ is $T = 1$, so any initial condition converges, under (2.23), to a fixed point. ∎

As we shall see in the examples in section (2.6), the fixed points of sequential iterations are not unique. On the other hand, it is an easy exercise to prove that the fixed points of synchronous and sequential updatings of Neural Networks are the same. Then:

Proposition 2.1. Let A be a symmetric matrix with non-negative diagonal entries; then the limit orbits of the sequential iteration of a Neural Network are only fixed points which are the same as the fixed points of the synchronous iteration of it.

Proof. It is deduced from Theorems 2.1, 2.2, and previous comment. ∎

Remarks.
1. If A is non symmetric or its diagonal, diag A, is not ≥ 0 we may obtain cycles of non-bounded periods. For instance take:

$$A = \begin{pmatrix} -1 & 1 & 0 & \cdots & 0 & 0 & 1 \\ 1 & -1 & 1 & \cdots & 0 & 0 & 0 \\ \vdots & \vdots & \vdots & & \vdots & \vdots & \vdots \\ 0 & 0 & 0 & \cdots & 1 & -1 & 1 \\ 1 & 0 & 0 & \cdots & 0 & 1 & -1 \end{pmatrix}$$

A is symmetric, diag $A = (-1, ..., -1)$. Take the null threshold vector $b = 0$.

Make the network evolve in a sequential form with the order $1 < 2 < ... < n$. Then the initial condition $(1, 0, ..., 0, 1) \in \{0, 1\}^n$ belongs to the following $n - 1$ cycle: $(1, 0, ..., 0, 1) \to (0, 0, ..., 1, 0) \to ... (0, 1, ..., 0, 0) \to (1, 0, ..., 0, 1)$.
2. If we consider the example given in remark 2 of section 2.4, and we iterate it in a sequential way with the following order of sites: $n < n - 1 < ... < 1$, the initial condition $(0, 1, ..., 1, 0) \in \{0, 1\}^n$ belongs to an $n - 1$ cycle.
3. Consider the two dimensional Bounded Neural Network (BNN), $\mathcal{A} = (I \times I, Q = \{0, 1\}, V_0^M, f)$ with zero boundary condition and the sequential update of f:

$$x_{(i,i')}(t + 1) =$$
$$\mathbb{1}\left(-x_{(i-1,i'-1)}(t+1) - x_{(i-1,i')}(t+1) - x_{(i-1,i'+1)}(t+1) - x_{(i,i'-1)}(t+1)\right.$$
$$\left. - 2.5x_{(i,i')}(t) - x_{(i,i'+1)}(t) - x_{i+1,i'-1}(t) - x_{(i+1,i'+1)}(t) + 3\right)$$

Clearly, matrix A is symmetric but its diagonal is not positive. In this context we observe large transient times and/or large periodic behaviour (see Figure 2.1).

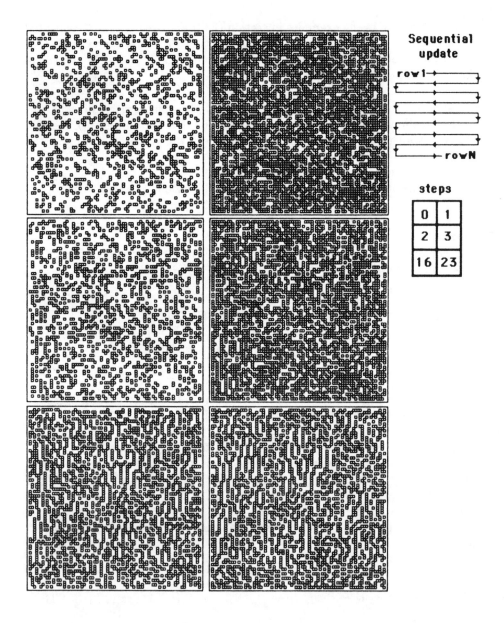

Figure 2.1. Sequential dynamics of a BNN with negative weights.

2.6. Block Sequential Iteration on Neural Networks

Let $I_1, ..., I_p$ be a partition of $I = \{1, ..., n\}$ ($I_r \neq \phi$ for any $r = 1, ..., p$, $I_r \cap I_{r'} = \phi$ for $r \neq r'$ and $\bigcup_{r=1}^{p} I_r = I$). By reordering I we can always assume $i < i'$ if $i \in I_r$, $i' \in I_{r'}$ with $r < r'$, and so $I_1, ..., I_p$ is an ordered partition with respect to the order of $\mathbb{Z}$.

We define the *block sequential iteration* as being sequential with respect to the order of blocks and synchronous within each block, more precisely:

$$\text{if} \quad i \in I_r : x_i(t+1) = \mathbb{1}\Big(\sum_{\substack{j \in \bigcup I_q \\ q < r}} a_{ij} x_j(t+1) + \sum_{\substack{j \in \bigcup I_q \\ q \geq r}} a_{ij} x_j(t) - b_i \Big)$$

$$\text{for} \quad t \geq 0 \tag{2.32}$$

The synchronous iteration is obtained when the partition is the trivial one: $p = 1$, $I_1 = I$; and the sequential iteration corresponds to the discrete partition: $p = n$, $I_r = \{r\}$ for $r \in I$.

Denote by T the period of the network iterated in a block sequential way with the partition $I_1, ..., I_p$. Let $X = (x(t) : t \in \mathbb{Z}_T)$ be a T-cycle for the evolution, i.e. $x(t)$ satisfies equation (2.32) for any $t \in \mathbb{Z}_T$. Denote by $X_i = (x_i(t) : t \in \mathbb{Z}_T)$ the T-cycle at the i-th site, and by $\gamma(X_i)$ the period of X_i. A useful functional for understanding the limit behavior of this iteration is again a difference of covariances among cycles weighted by the interactions, to wit:

$$L_b(X_i, X_j) = \begin{cases} a_{ij} \Delta V^{0,1}(X_i, X_j) & \text{if } i \in I_r, j \in I_q, r < q \\ a_{ij} \Delta V^{1,1}(X_i, X_j) & \text{if } i, j \in I_r \\ a_{ij} \Delta V^{1,0}(X_i, X_j) & \text{if } i \in I_r, j \in I_q, r > q \end{cases} \tag{2.33}$$

which can also be written:

$$L_b(X_i, X_j) = \begin{cases} \frac{a_{ij}}{T} \sum_{t \in \mathbb{Z}_T} x_i(t)(x_j(t) - x_j(t-1)) & \text{if } i \in I_r, j \in I_q, r < q \\ \frac{a_{ij}}{T} \sum_{t \in \mathbb{Z}_T} x_i(t)(x_j(t+1) - x_j(t-1)) & \text{if } i, j \in I_r \\ \frac{a_{ij}}{T} \sum_{t \in \mathbb{Z}_T} x_i(t)(x_j(t+1) - x_j(t)) & \text{if } i \in I_r, j \in I_q, r > q \end{cases} \tag{2.34}$$

From results of section 2.3 we deduce the following two properties which hold for general evolution equations $x(t+1) = F(x(t))$ or

$$x(t+1) = \bar{F}(x(t), ..., x(t-(r-1))).$$

Equality (2.7) implies:

$$\text{if } a_{ij} = a_{ji} \quad \text{then } L_b(X_i, X_j) + L_b(X_j, X_i) = 0 \tag{2.35}$$

and from lemma 2.2 we get:

$$\gamma(X_i) = 1 \quad \text{implies } L_b(X_i, X_j) = 0 \text{ for any } j \in I \tag{2.36}$$

A more precise result which takes into account the form (2.32) of the iteration is:

Lemma 2.5. Let $I_1, ..., I_p$ be an ordered partition of I, and the interaction matrix A such that it is block diagonal dominant for this partition, i.e.:

$$\text{for } r = 1, ..., p \quad \text{and any } i \in I_r : a_{ii} \geq \sum_{\substack{j \in I_r \\ j \neq i}} |a_{ij}|, \tag{2.37}$$

Then:

$$\sum_{j \in I} L_b(X_i, X_j) \leq 0 \quad \text{for any } i \in I \tag{2.38}$$

$$L_b(X_i, X_j) = 0 \text{ for any } j \in I \text{ iff } \sum_{j \in I} L_b(X_i, X_j) = 0 \text{ iff } \gamma(X_i) = 1 \tag{2.39}$$

$$\sum_{j \in I} L_b(X_i, X_j) < 0 \quad \text{iff } \gamma(X_i) > 1 \tag{2.40}$$

$$\sum_{i \in I} \sum_{j \in I} L_b(X_i, X_j) = 0 \quad \text{iff } \gamma(X_i) = 1 \quad \text{for any } i \in I \tag{2.41}$$

Proof. Equivalence (2.41) follows from (2.38) - (2.40). As before consider $\Gamma^0(X_i) = \{t \in \mathbb{Z}_T : x_i(t) = 0\}$, $\Gamma^1(X_i) = \{t \in \mathbb{Z}_T : x_i(t) = 1\}$ and $\varsigma(X_i)$, the class of maximal 1-chains in $\Gamma^1(X_i)$. Let us assume $\gamma(X_i) > 1$ so $\Gamma^1(X_i) \neq \mathbb{Z}_T$ and ϕ, which means that any maximal 1-chain is not 1-periodic. Let $\underline{t}_c$ be the unique element of C such that $\underline{t}_c - 1(\bmod T) \notin C$ and $\bar{t}_c = \underline{t}_c + (s_c - 1)(\bmod T)$ where s_c is the cardinality of C. Then $\bar{t}_c + 1 \notin C$.

Let $i \in I_r$, we have:

$$\sum_{j \in I} L_b(X_i, X_j) = \frac{1}{T} \sum_{C \in \varsigma(X_i)} \Big(\sum_{\substack{j \in \bigcup I_q \\ q < r}} a_{ij} (x_j(\bar{t}_c + 1) - x_j(\underline{t}_c)) +$$

$$\sum_{j \in I_r} a_{ij}(x_j(\bar{t}_c + 1) - x_j(\underline{t}_c - 1)) + \sum_{\substack{j \in \bigcup I_q \\ q > r}} a_{ij}(x_j(\bar{t}_c) - x_j(\underline{t}_c - 1)))$$

$$= \frac{1}{T} \sum_{C \in \varsigma(X_i)} \Big(\sum_{\substack{j \in \bigcup I_q \\ q < r}} a_{ij} x_j(\bar{t}_c + 1) + \sum_{\substack{j \in \bigcup I_q \\ q \geq r}} a_{ij} x_j(\bar{t}_c) - \sum_{\substack{j \in \bigcup I_q \\ q < r}} a_{ij} x_j(\underline{t}_c)$$

$$- \sum_{\substack{j \in \bigcup I_q \\ q \geq r}} a_{ij} x_j(\underline{t}_c - 1) + \sum_{j \in I_r} a_{ij}(x_j(\bar{t}_c + 1) - x_j(\bar{t}_c)))$$

Now $x_i(\bar{t}_c + 1) = 0$ so $\displaystyle \sum_{\substack{j \in \bigcup I_q \\ q < r}} a_{ij} x_j(\bar{t}_c + 1) + \sum_{\substack{j \in \bigcup I_q \\ q \geq r}} a_{ij} x_j(\bar{t}_c) < b_i$. On the

other hand $x_i(\underline{t}_c) = 1$ so $\displaystyle \sum_{\substack{j \in \bigcup I_q \\ q < r}} a_{ij} x_j(\underline{t}_c) + \sum_{\substack{j \in \bigcup I_q \\ q \geq r}} a_{ij} x_j(\underline{t}_c - 1) \geq b_i$. Then

$$\sum_{j \in I} L_b(X_i, X_j) < \frac{1}{T} \sum_{C \in \varsigma} \sum_{j \in I_r} a_{ij}(x_j(\bar{t}_c + 1) - x_j(\bar{t}_c)) =$$

$$\frac{1}{T} \sum_{C \in \varsigma} [(-a_{ii}) + \sum_{\substack{j \in I_r \\ j \neq i}} a_{ij}(x_j(\bar{t}_c + 1) - x_j(\bar{t}_c))]$$

The condition of block diagonal dominance (2.37) and the inequality

$$\Big| \sum_{\substack{j \in I_r \\ j \neq i}} a_{ij}(x_j(\bar{t}_c + 1) - x_j(\bar{t}_c)) \Big| \leq \sum_{\substack{j \in I_r \\ j \neq i}} |a_{ij}|, \text{ imply } \sum_{j \in I} L_b(X_i, X_j) < 0 \quad \blacksquare$$

Suppose that the matrix A is block diagonal dominant (i.e. satisfies (2.37)). Equivalence (2.41) expresses that $\sum_{i \in I} \sum_{j \in I} L_b(X_i, X_j) = 0$ for any cycle X is a necessary and sufficient condition to have $T = 1$. From (2.35) we deduce that A symmetric is a sufficient condition to have $\sum_{i \in I} \sum_{j \in I} L_b(X_i, X_j) = 0$ for any cycle X, so:

Theorem 2.3. [G7] Let $I_1, ..., I_p$ be a well ordered partition of I. Suppose the matrix of interactions A is symmetric and block diagonal dominant for the above partition. Then the block sequential iteration of the Neural Networks has period $T = 1$, that is, any limit cycle is a fixed point. ■

When we compare the fixed points of synchronous, sequential, and block sequential iterations we get:

Lemma 2.6. Let A be any iteration matrix, b a threshold vector, and $x \in \{0,1\}^n$. Then the property "x is a fixed point under block sequential iteration of an ordered partition $\alpha = (I_1, ..., I_p)$" does not depend on the partition α. Then the set of fixed points under any block sequential iteration is equal to the set of fixed points under synchronous or sequential iteration.

Proof. Let $\alpha = (I_1, ..., I_p)$ be an ordered partition and $x \in \{0,1\}^n$ be the initial condition of the block (α) sequential iteration and also of the synchronous iteration: $x^b(0) = x^{sy}(0) = x$. Denote by $x^b(1)$ (respectively $x^{sy}(1)$) the block (α) sequential iteration (respectively synchronous) iteration of x.

Assume $x^{sy}(1) = x$ so x is a fix point for synchronous iteration. By definition $x_i^b(1) = x_i^{se}(1)$ for any $i \in I_1$. By using recurrence arguments we show that $x^b(1) = x^{se}(1)$. Then x is also a fixed point under block sequential iteration. The converse is shown analogously. ■

Proposition 2.2. [G7] Let A be a symmetric matrix block diagonal dominant with respect to an ordered partition α. Then the limit orbits of the block (α) sequential iteration of a Neural Network are only fixed points which are the same as the fixed points of the synchronous iteration (or the sequential iteration) of it. ■

Note that if a matrix A is block diagonal dominant for a partition $\alpha = (I_1, ..., I_p)$ then it is also block diagonal dominant for a partition $\alpha' = (I_1', ..., I_{p'}')$ finer than α (recall that α' is finer than α, which we write as $\alpha' \geq \alpha$, iff any class I_s' of the partition α' is contained in some class I_r of α). Then we get:

Proposition 2.3. [G7] Let A be a symmetric matrix which is block diagonal dominant for the partition α. Then for any partition α' finer than α, the block sequential iteration of the Neural Network has only fixed points. ■

As a example, consider the lattice of partitions on $I = \{1, 2, 3, 4\}$, given in Figure 2.2.

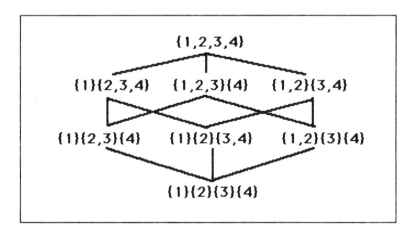

Figure 2.2. Partial order of partitions where $^{\alpha}\diagdown_{\alpha'}$ means α' is finer than α.

Take

$$A = \begin{pmatrix} 2 & 1 & -3 & 1 \\ 1 & 1 & 1 & -1 \\ -3 & 1 & 2 & 1 \\ 1 & -1 & 1 & 1 \end{pmatrix} \quad \text{and} \quad b = \begin{pmatrix} \frac{1}{2} \\ \frac{1}{2} \\ \frac{1}{2} \\ \frac{1}{2} \end{pmatrix}$$

For partitions $(\{1\}\{2, 3, 4\})$ and $(\{1, 2\}\{3, 4\})$ the symmetric matrix A is block diagonal dominant. Then A is also diagonal dominant for any partition finer than these, hence for all above partitions different from $\{1, 2, 3\}\{4\}$ and $\{1, 2, 3, 4\}$. Then Proposition 2.2 states that the block sequential iteration, with respect to any above partition, except $\{1, 2, 3, 4\}$ and $\{1, 2, 3\}\{4\}$, has as limit cycles only fixed points.

For any partition of the lattice we shall explicitly show the block sequential iteration that it induces, i.e. we shall give the iteration tree for all the vectors belonging to $\{0, 1\}^4$. When we write $x \to y$ we mean that y is the image of x under the block sequential iteration.

The different dynamics are given in Figure 2.3.

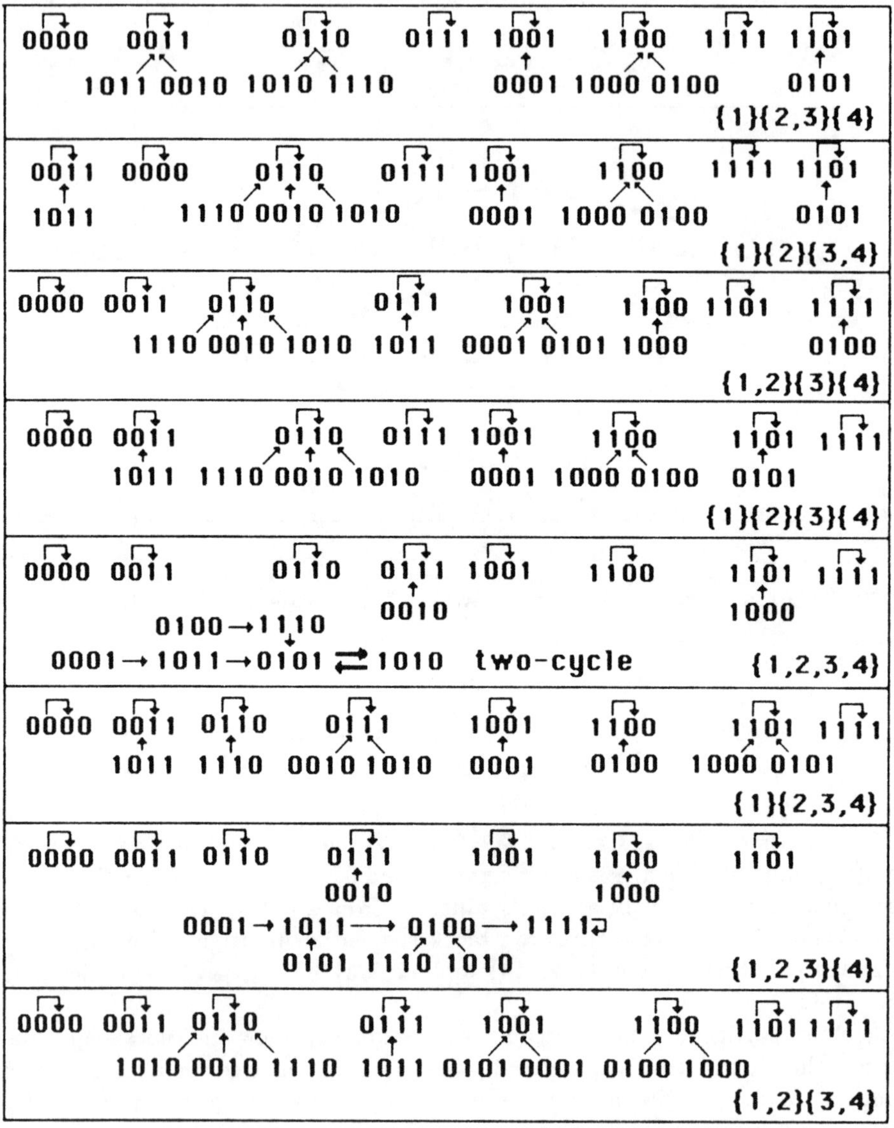

Figure 2.3. Block sequential dynamics for the partitions of $I = \{1, 2, 3, 4\}$.

2.7. Iteration with Memory

A mathematical model of Neural Networks proposed in [C] is represented by *k-memory iteration* which evolution equation is:

$$x_i(t+1) = 1\!\!1\Big(\sum_{j=1}^{n}\sum_{s=1}^{k} a_{ij}(s)x_j(t+1-s) - b_i\Big), \quad t \geq k-1 \qquad (2.42)$$

where $x_j(t) \in \{0,1\}$ for any $j \in I$. Note that the case $k = 1$ is just the synchronous iteration in section 2.4.

The dynamics generated by these equations have been studied for some particular one-dimensional systems, i.e. when there is only one site:

$$y(t) = 1\!\!1\Big(\sum_{s=1}^{k} a(s)y(t-s) - b\Big), \quad y(t) \in \{0,1\} \qquad (2.43)$$

Partial results in the study of evolution (2.43) under restrictive assumptions on coupling coefficients $a(s)$ can be seen in [CG,CMG,Ki,NS].

General results for the evolution (2.42) do not exist. In this paragraph we shall show some dynamical results when non-trivial regularities on coupling coefficients are satisfied. Our approach, just as in the other sections of this chapter, will consist of defining appropiate algebraic invariants.

Let T be the period of the Neural Network for k-memory evolution. Let $X = (x(t) : t \in \mathbb{Z}_T)$ be a periodic orbit, so $x(t)$ satisfies (2.42) for any $t \in \mathbb{Z}_T$. Denote by $X_i = (x_i(t) : t \in \mathbb{Z}_T)$ the T-cycle at site i and by $\gamma(X_i)$ its period. The functional defined on pairs of limit cycles comprises differences of covariances useful for studying the $k+1$ periodicity, conveniently weighted by the interactions. Let X_i, X_j be a pair of T-cycles, define:

$$L_{me}(X_i, X_j) = \sum_{s=1}^{k} a_{ij}(s)\Delta V^{k+1-s,s}(X_i, X_j) \qquad (2.44)$$

or more explicitly:

$$L_{me}(X_i, X_j) = \frac{1}{T} \sum_{t \in \mathbb{Z}_T} x_i(t)\Big\{\sum_{s=1}^{k} a_{ij}(s)(x_j(t+k+1-s) - x_j(t-s))\Big\} \qquad (2.45)$$

From lemma 2.2 we obtain the following property which is verified by any discrete system $x(t+1) = \bar{F}(x(t), ..., x(t-(r-1)))$ (and not only for (2.42)):

$$\text{if } \gamma(X_i)|k+1 \text{ then } L_{me}(X_i, X_j) = 0 \text{ for any } j \in I \qquad (2.46)$$

Now for evolution equation (2.42) we get:

Lemma 2.7. For any family of interaction matrices $(A(s) : s = 1, ..., k)$ we have:

$$\sum_{j \in I} L_{m\,e}(X_i, X_j) \leq 0 \text{ for any } i \in I \tag{2.47}$$

$$L_{m\,e}(X_i, X_j) = 0 \text{ for any } j \in I \text{ iff } \sum_{j \in I} L_{m\,e}(X_i, X_j) = 0 \text{ iff } \gamma(X_i)|k+1 \tag{2.48}$$

$$\sum_{j \in I} L_{m\,e}(X_i, X_j) < 0 \text{ iff } \gamma(X_i) \text{ does not divide } k+1 \tag{2.49}$$

$$\sum_{i \in I} \sum_{j \in I} L_{m\,e}(X_i, X_j) = 0 \text{ iff } \gamma(X_i)|k+1 \text{ for any } i \in I \tag{2.50}$$

Proof. (2.47) - (2.49) imply (2.50). Now consider $\Gamma^0(X_i) = \{t \in \mathbb{Z}_T : x_i(t) = 0\}$, $\Gamma^1(X_i) = \{t \in \mathbb{Z}_T : x_i(t) = 1\}$, $\varsigma(X_i)$ the class of maximal $k+1$-chains included in $\Gamma^1(X_i)$, and $\varsigma^{(k+1)}(X_i)$ the subclass of maximal $k+1$-chains which are $k+1$ periodic. As before, if $C \in \varsigma^{(k+1)}(X_i)$ we choose one of its elements $\underline{t}_c \in C$ and if $C \in \varsigma(X_i) \setminus \varsigma^{(k+1)}(X_i)$ we note by $\underline{t}_c$ the unique element in C such that $\underline{t}_c - (k+1)(\bmod T) \notin C$. Finally we write $\bar{t}_c = \underline{t}_c + (k+1)(s_c - 1) \pmod{T}$, where s_c is the cardinality of C.

From equality (2.10):

$$\sum_{j \in I} L_{m\,e}(X_i, X_j) = \frac{1}{T} \sum_{C \in \varsigma(X_i) \setminus \varsigma^{(k+1)}(X_i)} \sum_{j \in I} \sum_{s=1}^{k} a_{ij}(s)(x_j(\bar{t}_c + k + 1 - s) - x_j(\underline{t}_c - s)).$$

For any $C \in \varsigma(X_i) \setminus \varsigma^{(k+1)}(X_i)$ we have $\bar{t}_c + k + 1(\bmod T) \notin C$ so $x_i(\bar{t}_c + k + 1) = 0$ which is equivalent to $\sum_{j \in I} \sum_{s=1}^{k} a_{ij}(s)x_j(\bar{t}_c + k + 1 - s) < b_i$. On the other hand $x_i(\underline{t}_c) = 1$ so $\sum_{j \in I} \sum_{s=1}^{k} a_{ij}(s)x_j(\underline{t}_c - s) \geq b_i$. Then for any $C \in \varsigma(X_i) \setminus \varsigma^{(k+1)}(X_i)$ the sum $\sum_{j \in I} \sum_{s=1}^{k} a_{ij}(s)(x_j(\bar{t}_c + k + 1 - s) - x_j(\underline{t}_c - s))$ is < 0. As $\gamma(X_i)$ does not divide $k + 1$ the set $\varsigma(X_i) \setminus \varsigma^{(k+1)}(X_i)$ is non-empty so $\sum_{j \in I} L_{m\,e}(X_i, X_j) < 0$. ∎

Now assume that the set of interaction matrices $(A(s) : s = 1, ..., k)$ satisfy the palindromic condition:

$$A(k + 1 - s) = {}^t A(s) \text{ for } s = 1, ..., k \tag{2.51}$$

For any discrete evolution $x(t+1) = \bar{F}(x(t), ..., x(t-(r-1)))$ (i.e., form (2.42) is not necessary) we get the property:

$$
\begin{gathered}
\text{if } a_{ij}(k+1-s) = a_{ij}(s) \text{ for } s = 1, ..., k \text{ then} \\
L_{me}(X_i, X_j) + L_{me}(X_j, X_i) = 0 \quad \forall i, j \in I.
\end{gathered}
\tag{2.52}
$$

This assertion follows readily from (2.7), in fact: $L_{me}(X_i, X_j) =$

$$
- \sum_{s=1}^{k} a_{ij}(s) \Delta V^{s, k+1-s}(X_j, X_i) = - \sum_{s=1}^{k} a_{ij}(k+1-s) \Delta V^{s, k+1-s}(X_j, X_i) =
$$

$$
- \sum_{s=1}^{k} a_{ij}(s') \Delta V^{k+1-s', s'}(X_j, X_i) = -L_{me}(X_j, X_i).
$$

Now equivalence (2.50) means that the condition $\sum_{i \in I} \sum_{j \in I} L_{me}(X_i, X_j) = 0$ for any cycle X is a necessary and sufficient one for $T | k+1$. From (2.52) we deduce that the palindromic condition (2.51) of the sequence of matrices $(A(s) : s = 1, ..., k)$ is a sufficient condition for $\sum_{i \in I} \sum_{j \in I} L_{me}(X_i, X_j) = 0$ for any cycle X, so:

Theorem 2.4. [G9,PT3] If the class of interaction matrices $(A(s) : 1 \leq s \leq k)$ is palindromic (i.e. satisfy (2.51)) then the period T of the Neural Network iterated with k-memory satisfies $T | k+1$. ∎

The above theorem also gives information on limit behavior of Neural Networks iterated with non-connected memory. More precisely, consider the evolution:

$$
x_i(t+1) = \mathbb{1}\left(\sum_{j \in I} \sum_{l=1}^{p} a_{ij}(l) x_j(t+1-r_l) - b_i\right)
\tag{2.53}
$$

where the integral memory steps satisfy: $r_p > r_{p-1} > ... > r_1 \geq 1$. Now assume that this steps satisfy the following symmetrical condition with respect to the half-middle point $\frac{r_1 + r_p}{2}$:

$$
r_l + r_{p-l+1} = r_1 + r_p \text{ for any } l = 1, ..., p
\tag{2.54}
$$

(for instance this occurs if $\{r_1, ..., r_p\} = \{r, r+1, ..., r+p-2, r+p-1\}$ are consecutive integers).

Let $k = r_1 + r_p - 1$ define the family of matrices $(D(s) : s = 1, ..., k)$ with

$$
D(s) = \begin{cases} A(l) & \text{if } s = r_l \\ 0 & \text{if } s \notin \{r_1, ..., r_p\} \end{cases} \quad \text{i.e. } d_{ij}(s) = \sum_{l=1}^{p} a_{ij}(l) \delta_{s, r_l}
\tag{2.55}
$$

(where $\delta_{s,s'}$ is the Kroenecker delta function).

Then iteration (2.53) can be written as:

$$x_i(t+1) = \mathbb{1}\left(\sum_{j=1}^{n}\sum_{s=1}^{k} d_{ij}(s)x_j(t+1-s) - b_i\right) \tag{2.56}$$

If the initial set of matrices $(A(l) : l = 1, ..., p)$ satisfy the following palindromic condition:

$$A(p+1-l) = {}^t A(l) \quad \text{for any } l = 1, ..., p \tag{2.57}$$

then the family of matrices $(D(s) : s = 1, ..., k)$ satisfy the palindromic condition (2.51) i.e.:

$$D(k+1-s) = {}^t D(s) \quad \text{for } s = 1, ..., k \tag{2.58}$$

By the above consideration we obtain the following generalization of Theorem 2.4:

Proposition 2.4. [G9] Let T be the period of a Neural Network whose iteration is given by (2.53) with memory steps $r_p > r_{p-1} ... > r_1 \geq 1$ satisfying the symmetric condition (2.54). If the class of matrices $(A(l) : l = 1, ..., p)$ satisfy the palindromic condition (2.57) then $T | r_1 + r_p$.

Proof. It is direct from Theorem 2.4 because (2.58) is satisfied. With the above notation $k + 1 = r_1 + r_p$. ∎

Remarks.

1. Consider evolution (2.53). By using its equivalent form (2.56) we obtain the following functional from (2.44): $L(X_i, X_j) = \sum_{l=1}^{p} a_{ij}(l)\Delta V^{r_1 + r_p - r_l, r_l}(X_i, X_j) = \frac{1}{T}\sum_{t \in \mathbb{Z}_T} x_i(t)(\sum_{l=1}^{p} a_{ij}(r_l)(x_j(t + r_1 + r_p - r_l) - x_j(t - r_l)))$ which satisfies properties entirely analogous to those of L_{me}.

2. If $A(l) = A$ for any $l = 1, ..., p$ the condition (2.57) is reduced to the symmetric condition $A = {}^t A$. If $k = 1$ we find exactly the synchronous iteration of section 2.4. Theorem 2.4 asserts that its period divides $k + 1 = 2$, which is the same conclusion as that in Theorem 2.1.

3. An obvious case where the palindromic condition is satisfied occurs when the interactions are constant $a_{ij}(s) = a$.

4. Consider the following one-site example, with non-connected memory:

$$x(t+1) = 1\left(-\sum_{l=r}^{p+r-1} x(t+1-l) + \frac{1}{2}\right), \quad r \geq 1$$

We have that the integer memory steps are $\{r, r+1, ..., r+p-1\}$, and $a(l) = -1$ for any $l = 1, ..., p$ (see (2.53)). Then the hypotheses of Proposition 2.4 are fullfilled, so its period T satisfies $T|2r+p-1$ (recall that $r_1 + r_p = 2r+p-1$). Moreover, for any $r \geq 1$ this iteration admits the vector $(1^r 0^{r+p-1})$ as a cycle of length $2r+p-1$ ($1^r 0^{r+p-1}$ means the vector with the first r coordinates '1' and the other $r+p-1$ equal to '0'). Hence for any $r \geq 1$, $p \geq 1$ the maximum period $r_1 + r_p$ is attained.

Furthermore, given $k \geq 1$, and $q|k+1$ we can construct a one-site iteration, with k-memory, possesing a limit cycle of period q. For $q = 1$ it is evident that we can construct such iteration and the above example shows that the case $q = k+1$ can also be constructed. Then assume $1 < q < k+1$. Take $(a(s) : s = 1, ..., k) = (0^{q-1} 1 0^{k-2q} 1 0^{q-1})$ which satisfies the palindromic condition (2.51). Consider the following k-memory evolution:

$$x(t+1) = 1\left(\sum_{s=1}^{k} a(s)x(t+1-s) - 2\right)$$

It is easy to show that this iteration admits the q-cycle: $(0^{q-1} 1 0^{q-1} 1...0^{q-1} 1)$ (i.e. $0^{q-1} 1$ repeated $\frac{k+1}{q}$ times).

2.8. Synchronous Iteration on Majority Networks

The Automata Network here studied represents the evolution in time of the opinion of the people in a society. In this model each person makes its choice according to a *majority rule* which takes into account the influences of the other members. In a tie case the choice of a person is made according to his own hierarchy of these opinions. Any opinion is also weighted by a global social criterium.

More precisely, let $I = \{1, ..., n\}$ be the members of the society, $Q = \{0, ..., p-1\}$ be the set of opinions 'which is ordered by the same order of $\mathbb{R}$', each opinion q being weighted by a social factor $\alpha(q) \in \mathbb{R}$. The hierarchy of the opinions of person i is expressed by a permutation σ_i of Q into itself: $\sigma_i(q) > \sigma_i(q')$ means the person i prefers opinion q to q'. The interactions or influences among the opinions of members is given by a matrix $A = (a_{ij} : i,j \in I) : a_{ij} > 0$ means friendly, $a_{ij} = 0$ indifferent, $a_{ij} < 0$ unfriendly.

Let $z(t) = (z_i(t) : i \in I) \in Q^n$ be the configuration of the society opinion at time t. Its evolution is given by:

$$z_i(t+1) = q \text{ iff } \begin{cases} \alpha(q) \sum\limits_{j \in I: z_j(t)=q} a_{ij} \geq \alpha(q') \sum\limits_{j \in I: z_j(t)=q'} a_{ij} \text{ if } \sigma_i(q) \geq \sigma_i(q') \\ \alpha(q) \sum\limits_{j \in I: z_j(t)=q} a_{ij} > \alpha(q') \sum\limits_{j \in I: z_j(t)=q'} a_{ij} \text{ if } \sigma_i(q) < \sigma_i(q') \end{cases}$$

$$(2.59)$$

For notation we code any $q \in Q$ as a p-sequence $\hat{q} = (0, ..., 1, ..., 0)$ where the '1' is at the q-th place. Then the configuration space Q^n is equivalent to $S = \{x = ((x_i^q : q \in Q) : i \in I) \in ((\{0,1\}^p))^n : \sum\limits_{q \in Q} x_i^q = 1\}$. The state $z = (z_i : i \in I) \in Q^n$ is thus coded as $x = ((x_i^q : q \in Q) : i \in I) \in S$ in such a way that: $z_i = q$ iff $x_i^q = 1$ and $x_i^{q'} = 0$ for any $q' \neq q$. The evolution (2.59) is written in the state space S as:

$$x_i^q(t+1) = 1 \text{ iff } \begin{cases} \alpha(q) \sum\limits_{j \in I} a_{ij} x_j^q(t) \geq \alpha(q') \sum\limits_{j \in I} a_{ij} x_j^{q'}(t) \text{ if } \sigma_i(q) \geq \sigma_i(q') \\ \alpha(q) \sum\limits_{j \in I} a_{ij} x_j^q(t) > \alpha(q') \sum\limits_{j \in I} a_{ij} x_j^{q'}(t) \text{ if } \sigma_i(q) < \sigma_i(q') \end{cases}$$

$$(2.60)$$

By the finiteness of the state space, any initial condition evolves in finite time to a finite cycle. Let T be the period of the above system, $X = (x(t) : t \in \mathbb{Z}_T)$ a limit T-cycle, $X_i = (x_i(t) : t \in \mathbb{Z}_T)$ the local cycle on the i-th site, and $\gamma(X_i)$ the period of X_i. Let $X_i^q = (x_i^q(t) : t \in \mathbb{Z}_T)$, if $\gamma(X_i^q)$ is its period we necessarily have $\gamma(X_i^q) | \gamma(X_i)$.

The algebraic invariant will be a weighted sum between differences of covariances at a single step, more precisely:

$$L_{Ma}(X_i, X_j) = a_{ij} \sum_{q \in Q} \alpha(q) \Delta V^{1,1}(X_i^q, X_j^q) \qquad (2.61)$$

which can be developped as:

$$L_{Ma}(X_i, X_j) = \frac{a_{ij}}{T} \sum_{q \in Q} \alpha(q) \sum_{t \in \mathbb{Z}_T} x_i^q(t)(x_j^q(t+1) - x_j^q(t-1)) \qquad (2.62)$$

As before, we can obtain some results which hold for any evolution equation $x(t+1) = F(x(t))$, or $x(t+1) = \bar{F}(x(t), ..., x(t-(r-1)))$, in S. From (2.7):

$$a_{ij} = a_{ji} \quad \text{implies } L_{Ma}(X_i, X_j) + L_{Ma}(X_j, X_i) = 0 \qquad (2.63)$$

and lemma 2.2 and relation $\gamma(X_i^q)|\gamma(X_i)$ imply:

$$\text{if } \gamma(X_i)|2 \quad \text{then } L_{Ma}(X_i, X_j) = 0 \tag{2.64}$$

When we use the particular form of equation (2.60) we deduce more precise relationship between the periods and the functional.

Lemma 2.8. For any interaction matrix A:

$$\sum_{j \in I} L_{Ma}(X_i, X_j) \leq 0 \quad \text{for any } i \in I. \tag{2.65}$$

$$L_{Ma}(X_i, X_j) = 0 \ \forall j \in I \text{ iff } \sum_{j \in I} L_{Ma}(X_i, X_j) = 0 \text{ iff } \gamma(X_i)|2 \tag{2.66}$$

$$\sum_{j \in I} L_{Ma}(X_i, X_j) = 0 \text{ iff } \gamma(X_i) > 2 \tag{2.67}$$

$$\sum_{i \in I} \sum_{j \in I} L_{Ma}(X_i, X_j) = 0 \text{ iff } \gamma(X_i)|2 \text{ for any } i \in I \tag{2.68}$$

Proof. Property (2.68) follows from (2.65) - (2.67). Let $\Gamma^0(X_i^q) = \{t \in \mathbb{Z}_T : x_i^q(t) = 0\}$, $\Gamma^1(X_i^q) = \{t \in \mathbb{Z}_T : x_i^q(t) = 1\}$, $\varsigma(X_i^q)$ be the class of maximal 2-chains of $\Gamma^1(X_i^q)$ and $\varsigma^{(2)}(X_i^q)$ be the subclass of 2-periodic maximal 2-chains. Also choose t_c, $\bar{t}_c = t_c + 2(s_c - 1)$ such that $C = \{t_c + 2l : 0 \leq l < s_c\}$. Thus $C \in \varsigma(X_i^q) \setminus \varsigma^{(2)}(X_i^q)$ iff $\bar{t}_c + 2(\bmod T) \notin C$. From (2.10) we can write:

$$\sum_{j \in I} L_{Ma}(X_i, X_j) = \frac{1}{T} \sum_{q \in Q} \sum_{C \in \varsigma(X_i^q) \setminus \varsigma^{(2)}(X_i^q)} \alpha(q) \sum_{j \in I} a_{ij} (x_j^q(\bar{t}_c + 1) - x_j^q(\bar{t}_c - 1)) \tag{2.69}$$

By definition the supports $\{\Gamma^1(X_i^q) : q \in Q\}$ are mutually disjoint. Let us denote $\mathcal{D} = \{C \in \bigcup_{q \in Q} \varsigma(X_i^q) \setminus \varsigma^{(2)}(X_i^q)\}$, then if $C \neq C'$ are elements of $\mathcal{D}$ we necesarily have $C \cap C' = \phi$. Then the points $\{\bar{t}_c + 1(\bmod T) : C \in \mathcal{D}\}$ are different from one another, as well as from $\{\bar{t}_c - 1(\bmod T) : C \in \mathcal{D}\}$. Denote $\bar{G} = \{\bar{t}_c + 1(\bmod T) : C \in \mathcal{D}\}$, $\underline{G} = \{\underline{t}_c - 1(\bmod T) : C \in \mathcal{D}\}$. We shall prove $\bar{G} = \underline{G}$.

Let $\bar{t}_c \in C$ with $C \in \varsigma(X_i^q) \setminus \varsigma^{(2)}(X_i^q)$. By definition $\bar{t}_c + 2(\bmod T) \notin C$, then there exists some $q' \neq q$ such that $\bar{t}_c + 2(\bmod T) \in C'$ with $C' \in \varsigma(X_i^{q'})$. As $(\bar{t}_c + 2) - 2(\bmod T) \notin C'$ we necessarily have $C' \in \varsigma(X_i^{q'}) \setminus \varsigma^{(2)}(X_i^{q'})$.

So $\bar{t}_c + 2(\bmod T) = \underline{t}_{c'}$. Then $\bar{t}_c + 1(\bmod T) = \underline{t}_{c'} - 1$. Hence the inclusion $\bar{G} \subset \underline{G}$ holds true; in an analogous way we show $\underline{G} \subset \bar{G}$.

Now for $C \in \mathcal{D}$, we denote by $q(C)$ the unique element in Q such that $C \in \varsigma(X_i^q)$. Pick $t \in \bar{G} = \underline{G}$. We write $\bar{q}(t) = q(C)$ if $t = \bar{t}_c + 1(\bmod T)$, and $\underline{q}(t) = q(C')$ if $t = \underline{t}_{c'} - 1$. Obviously $\bar{q}(t) \neq \underline{q}(t)$.

Let us develop (2.69) by using the above remarks and notations:

$$\sum_{j \in I} L_{M\,a}(X_i, X_j) =$$

$$\frac{1}{T} \sum_{C \in \mathcal{D}} \left(\alpha(q(C)) \sum_{j \in I} a_{ij} x_j^{q(c)}(\bar{t}_c + 1) - \alpha(q(C)) \sum_{j \in I} a_{ij} x_j^{q(c)}(\underline{t}_c - 1) \right)$$

$$= \frac{1}{T} \left(\sum_{t \in \bar{G}} \alpha(\bar{q}(t)) \sum_{j \in I} a_{ij} x_j^{\bar{q}(t)}(t) - \sum_{t \in \underline{G}} \alpha(\underline{q}(t)) \sum_{j \in I} a_{ij} x_j^{q(t)}(t) \right)$$

$$= \frac{1}{T} \sum_{t \in \bar{G}} \left(\alpha(\bar{q}(t)) \sum_{j \in I} a_{ij} x_j^{\bar{q}(t)}(t) - \alpha(\underline{q}(t)) \sum_{j \in I} a_{ij} x_j^{q(t)}(t) \right)$$

If $C \in \varsigma(X_i^q) \setminus \varsigma^{(2)}(X_i^q)$ we have $x_i^q(\underline{t}_c) = 1$. Evolution equation (2.60) implies that, in terms of $t \in \underline{G} = \bar{G}$, this equality is equivalent to the relations (2.70), (2.71) below:

$$\alpha(\underline{q}(t)) \sum_{j \in I} a_{ij} x_j^{q(t)}(t) = \max_{q' \in Q} \alpha(q') \sum_{j \in I} a_{ij} x_j^{q'}(t) \tag{2.70}$$

and $\alpha(\underline{q}(t)) \displaystyle\sum_{j \in I} a_{ij} x_j^{q(t)}(t) > \max_{q' \in Q\,:\,\sigma_i(q') > \sigma_i(\underline{q}(t))} \alpha(q') \displaystyle\sum_{j \in I} a_{ij} x_j^{q'}(t)$ (2.71)

In particular $\alpha(\underline{q}(t)) \displaystyle\sum_{j \in I} a_{ij} x_j^{q(t)}(t) \geq \alpha(\bar{q}(t)) \displaystyle\sum_{j \in I} a_{ij} x_j^{q(t)}(t)$. Then $\displaystyle\sum_{j \in I} L_{M\,a}(X_i, X_j) \leq 0$. Let us prove that the strict inequality $\displaystyle\sum_{j \in I} L_{M\,a}(X_i, X_j) < 0$ actually holds.

Let $Q_i' = \{q \in Q : \gamma(X_i^q) > 2\}$ be the set of q whose cycles X_i^q are of period strictly greater than two. As $\gamma(X_i) > 2$ we have $Q_i' \neq \phi$, furthermore it can be easily shown that the cardinality of Q_i' is ≥ 2 (if only one $q \in Q_i'$ existed, we would arrive at a contradiction because $\gamma(X_i^q) \leq 2$ for any $q' \neq q$ implies that $\gamma(X_i^q) \leq 2$).

Take q_i' such that $\sigma_i(q_i') = \inf\{\sigma_i(q) : q \in Q_i'\}$. Let $C \in \varsigma(X_i^{q_i'}) \setminus \varsigma^{(2)}(X_i^{q_i'})$ and $t^* = \underline{t}_c - 1 \in \underline{G}$. We have $\underline{q}(t^*) = q_i'$. Now $\bar{q}(t^*) \neq \underline{q}(t^*)$, but $\bar{q}(t)$ also belongs to Q_i'; then $\sigma_i(\bar{q}(t^*)) > \sigma_i(\underline{q}(t^*))$. Condition (2.71) implies $\alpha(\underline{q}(t^*)) \displaystyle\sum_{j \in I} a_{ij} x_j^{q(t^*)}(t^*) > \alpha(\bar{q}(t^*)) \displaystyle\sum_{j \in I} a_{ij} x_j^{q(t^*)}(t^*)$. Hence $\displaystyle\sum_{j \in I} L_{M\,a}(X_i, X_j) < 0$. ∎

From equivalence (2.68) of lemma 2.8 it follows that the condition $\sum_{i \in I} \sum_{j \in I} L_{M\,a}(X_i, X_j) = 0$ for any cycle X is a necessary and sufficient condition for $T|2$. On the other hand, according to (2.63) the symmetry of A is a sufficient condition for $\sum_{i \in I} \sum_{j \in I} L_{M\,a}(X_i, X_j) = 0$ for any cycle X, so:

Theorem 2.5. [GT1] If A is symmetric then the period T of the Neural Network iterated by the majority rule (2.59) satisfies $T|2$. ∎

Remarks.

1. Previous theorem was also proved in [PS] by combinatorial arguments.

2. We can perform the same analysis as we did for threshold functions for k-memory majority networks. Hence, under a palindromic hypothesis we can prove that its limit period T satisfies $T|k + 1$.

References

[C1] Caianiello, E.R., *Decision Equations and Reverberations*, Kybernetik, 3(2), 1966.

[CG] Cosnard, M., E. Goles, *Dynamique d'un Automate à Mémoire Modélisant le Fonctionnement d'un Neurone*, C.R. Acad. Sc., 299(10), Série I, 1984, 459-461.

[CMG] Cosnard, M., D. Moumida, E. Goles, T. De St. Pierre, *Dynamical Behaviour of a Neural Automaton with Memory*, Complex Systems, 2, 1988, 161-176.

[G5] Goles, E., *Sequential Iterations of Threshold Functions*, Numerical Methods in the Study of Critical Phenomena, Delladora et al eds. Springer-Verlag, Series in Sygernetics, 1981, 64-70.

[G7] Goles. E., *Fixed Point Behaviour of Threshold Functions on a Finite Set*, SIAM J. on Alg. and Disc. Meths., 3(4), 1982, 529-531.

[G9] Goles. E., *Dynamical Behaviour of Neural Networks*, SIAM J. Disc. Alg. Meth., 6, 1985, 749-754.

[GO3] Goles, E., J. Olivos, *Comportement Pèriodique des Fonctions à Seuil Binaires et Applications*, Disc. App. Math., 3, 1981, 95-105.

[GT1] Goles, E., M. Tchuente, *Iterative Behaviour of Generalized Majority Functions*, Math. Soc. Sci., 4, 1984.

[Ki2] Kitagawa, T., *Dynamical Systems and Operators Associated with a Single Neuronic Equation*, Math. Bios., 18, 1973.

[NS] Nagumo, J., S. Sato, *On a Response Characteristic of a Mathematical Neuron Model*, Kybernetic, 3, 1972, 155-164.

[PS] Poljak, S., M. Sura, *On Periodical Behaviour in Society with Symmetric Influences*, Combinatorica, 3, 1983, 119-121.

[PT3] Poljak, S., D. Tursik, *On Systems, Periods and Semipositive Mappings*, Comm. Math., Univ. Carolinae 25, 4, 1984, 597-614.

3. LYAPUNOV FUNCTIONALS ASSOCIATED TO NEURAL NETWORKS

3.1. Introduction

In Chapter 2 we analized the steady state behavior of Neural and Majority Networks by means of algebraic invariants which enabled the characterization of periods. Unfortunately, the transient behavior does not yield as readily to a study based on such class of invariants. To overcome the difficulty, we introduce here Lyapunov functionals driving the network dynamics. Using this kind of functionals, explicit bounds to the transient length will be found. Moreover, Lyapunov functionals provide a physical interpretation of Neural Networks. These operators were first introduced by Hopfield [Ho2] to analize the fixed point behaviour of random sequential iterations of associative networks. As for their application in the study of synchronous updating rules and of memory updating, they were defined and developed in [G2,GFP]. Besides, they were applied to a reversible automaton in [Po].

3.2. Synchronous Iteration

For a dynamics $x(t+1) = F(x(t), x(t-1), ...)$ a real functional $E(x(t))$ is called a *Lyapunov functional* if it is decreasing: $E(x(t+1)) \leq E(x(t))$ for any $t \geq 1$. Hence if $\{x(t) : t \in \mathbb{Z}_T\}$ is a T-periodic orbit, then necessarily the functional is constant on it, i.e. $E(x(t)) = E_0$ for any $t \in \mathbb{Z}_T$. A Lyapunov functional is said to be strictly decreasing (or *strictly Lyapunov functional*) if $E(x(t+1)) < E(x(t))$ when $x(t)$ does not belong to a periodic orbit. When time is indexed by some other subset of $\mathbb{R}$, a similar definition can be given.

While a Lyapunov functional maps $\{0,1\}^n$ onto $\mathbb{R}$, that is $x \to E(x)$, in a number of particular cases we shall encounter explicit formulae for Lyapunov functionals depending on $x(t)$ and $x(t-1)$. We shall however overlook the fact since such functionals can always be put in the form $E(x(t))$.

Now consider the synchronous iteration on the finite Neural Network $\mathcal{N} = (I, A, b)$ where $I = \{1, ..., n\}$:

$$x(t+1) = \bar{\mathbb{1}}(Ax(t) - b) \quad x(t) \in \{0,1\}^n \qquad (3.1)$$

From the finiteness of the set $\{0,1\}^n$ we can assume that the threshold vector b satisfies:

$$(Au)_i = \sum_{j \in I} a_{ij} u_j \neq b_i \quad \forall i \in I, \quad \forall u \in \{0,1\}^n \qquad (3.2)$$

Let us show this assertion. Consider the finite set $C_i = \{a = (Au)_i : u \in \{0,1\}^n\}$ and define $\nu_i = \max\{a < b_i : a \in C_i\}$. Besides, any vector $\tilde{b} = (\tilde{b}_i : i \in I)$ satisfying $\tilde{b}_i \in (\nu_i, b_i)$ is such that the synchronous evolution $x(t+1) = \bar{\mathbb{1}}(Ax(t) - \tilde{b})$ is the same as the dynamics (3.1) with b instead of $\tilde{b}$. Furthermore the definition of $\tilde{b}$ implies $(Au)_i \neq \tilde{b}_i \; \forall i \in I$ and $\forall u \in \{0,1\}^n$, hence the assumption (3.2) can always be made. This fact will be important to obtain strictly decreasing properties of the functionals we shall introduce.

Now let $\{x(t) : t \geq 0\}$ be a trajectory of the synchronous iteration and define the following functional for $t \geq 1$:

$$E_{sy}(x(t)) = -\sum_{i \in I} x_i(t) \sum_{j \in I} a_{ij} x_j(t-1) + \sum_{i \in I} b_i(x_i(t) + x_i(t-1)) \qquad (3.3)$$

or, in a more compact form:

$$E_{sy}(x(t)) = - < x(t), Ax(t-1) > + < b, x(t) + x(t-1) > \qquad (3.4)$$

where $< x, y >= \sum_{i \in I} x_i y_i$ is the usual inner product.

Propositon 3.1. ([G2,GFP]). Let A be a symmetric real matrix; then the functional $E_{sy}(x(t))$ is a strictly decreasing Lyapunov functional for the synchronous iteration.

Proof. Since A is a symmetric matrix we have:

$$\Delta_t E_{sy} = E_{sy}(x(t)) - E_{sy}(x(t-1)) = -\sum_{i \in I}(x_i(t) - x_i(t-2))\left(\sum_{j \in I} a_{ij} x_j(t) - b_i\right) \quad (3.5)$$

From the definition of synchronous iteration, for any $i \in I$ we find:

$$-(x_i(t) - x_i(t-2))\left(\sum_{j \in I} a_{ij} x_j(t) - b_i\right) \leq 0.$$

So we conclude $\Delta_t E_{sy} \leq 0$ and $\Delta_t E_{sy} < 0$ iff $x_i(t) \neq x(t-2)$. ∎

Corollary 3.1. If A is symmetric the orbits of the synchronous iteration are only fixed points and/or two cycles.

Proof. Suppose $(x(0), ..., x(T-1))$ is a cycle of period T. Then $E_{sy}(x(t))$ is necesarily constant for $t = 0, ..., T-1$. If $T > 2$ we have $x(2) \neq x(0)$, so Proposition 3.1 implies $E_{sy}(x(2)) < E_{sy}(x(1))$ which is a contradiction. ∎

Hence we have shown in a very straightforward way the result given by Theorem 2.1. In fact it can be shown that for a symmetric matrix A the quantities E_{sy} and L_{sy} (see Chapter 2) are connected by a simple algebraic relation independent of the synchronous iteration. More precisely, let $Y = \{y(t) : t \in \mathbb{Z}\}$ be any sequence of vectors $y(t) \in \{0,1\}^n$. For any $T \geq 1$ define:

$$L_{sy}(Y) = \frac{1}{T}\sum_{t=0}^{T-1}\sum_{i\in I}\sum_{j\in I} a_{ij}y_i(t)(y_j(t+1) - y_j(t-1))$$

and consider $E_{sy}(y(t))$ for $t \geq 1$ as defined in (3.3). Then it is easy to show that the following equality holds:

$$L_{sy}(Y) = \frac{1}{T}(\sum_{t=0}^{T-1}(E_{sy}(y(t)) - E_{sy}(y(t-1)) + \sum_{t=0}^{T-1}\sum_{i\in I} b_i(y_i(t+1) - y_i(t-1))) \quad (3.6)$$

Obviously when $Y = (y(t) : t \in \mathbb{Z})$ is a T-cycle then the right-hand side of (3.6) vanishes.

To study the transient phase it is more convenient to work with another Lyapunov functional which is only slightly different from E_{sy}. Define:

$$\begin{aligned}E_{sy}^*(x(t)) = &-\sum_{t\in I}((2x_i(t) - 1)\sum_{j\in I} a_{ij}(2x_j(t-1) - 1)) \\ &+ \sum_{i\in I}(2b_i - \sum_{j\in I} a_{ij})(2x_i(t) - 1 + 2x_i(t-1) - 1)\end{aligned} \quad (3.7)$$

Proposition 3.2. Assume the matrix A to be symmetric. The difference $\Delta_t E_{sy}^* = E_{sy}^*(x(t)) - E_{sy}^*(x(t-1))$ satisfies $\Delta_t E_{sy}^* = 4\Delta_t E_{sy}$ so $E_{sy}^*(x(t))$ is a strictly decreasing Lyapunov functional for the synchronous iteration.

Proof. From the symmetry of A:

$$\begin{aligned}\Delta_t E_{sy}^* = &-\sum_{i\in I}((2x_i(t) - 2x_i(t-2))\sum_{j\in I} a_{ij}(2x_j(t-1) - 1)) \\ &+ \sum_{i\in I}(2b_i - \sum_{j\in I} a_{ij})(2x_i(t) - 2x_i(t-2)) \\ = &-2\sum_{i\in I}(x_i(t) - x_i(t-2))(2\sum_{j\in I} a_{ij}x_j(t-1) - \sum_{j\in I} a_{ij} - 2b_i + \sum_{j\in I} a_{ij})\end{aligned}$$

From expression (3.5) we find $\Delta_t E_{sy}^* = 4\Delta_t E_{sy}$, so Proposition 3.1 allows us to conclude the result. ∎

Since we are working in the finite set $\{0,1\}^n$ we can find bounds for the functional E_{sy}^*, as follows:

Proposition 3.3. Let $\{x(t) : t \le 0\}$ be a trajectory. Then $E_{sy}^*(x(t))$ is bounded by:

$$-\|A\|_1 - 2\|2b - A1\|_1 \le E^*(x(t)) \le -2\sum_{i \in I} e_i + \|2b - A\bar{1}\|_1 \qquad (3.8)$$

where:

$$e_i = \min\{|\sum_{j \in I} a_{ij} u_j - b_i| : u \in \{0,1\}^n\} \qquad (3.9)$$

$\|A\| = \sum_{i \in I}\sum_{j \in I} |a_{ij}|$, $\bar{1} = (1, ..., 1)$ is the 1-constant vector, and $\|u\|_1 = \sum_{i \in I} |u_i|$ for any vector $u \in \mathbb{R}^n$.

Proof. Consider the i-th term in expression (3.7) of $E^*(x(t))$, i.e.:

$$(E_{sy}^*(x(t)))_i = -(2x_i(t) - 1)(\sum_{j \in I} a_{ij}(2x_j(t-1) - 1))$$
$$+ (2b_i - \sum_{j \in I} a_{ij})(2x_i(t) - 1 + 2x_i(t-1) - 1)$$

Hence the following lower bound is obtained:

$$(E_{sy}^*(x(t)))_i \ge -\sum_{j \in I} |a_{ij}| - 2|2b_i - \sum_{j \in I} a_{ij}|$$

To obtain an upper bound write $(E_{sy}^*(x(t)))_i$ as follows:

$$(E_{sy}^*(x(t)))_i = -2(2x_i(t) - 1)(\sum_{j \in I} a_{ij}x_j(t-1) - b_i) + (2b_i - \sum_{j \in I} a_{ij})(2x_i(t-1) - 1)$$

From the definition of the synchronous iteration, the first term of the right hand side is always non-positive, in fact it is bounded by $-2e_i$. Hence for any trajectory $x(t)$ we get:

$$(E_{sy}^*(x(t)))_i \le -2e_i + |2b_i - \sum_{j \in I} a_{ij}|$$

Since $E_{sy}^*(x(t)) = \sum_{i \in I}(E_{sy}^*(x(t)))_i$, we get (3.8). ∎

Before giving a bound for the transient length, let us introduce some notation. Let $\{x(t) : t \geq 0\}$ be the trajectory starting from $x(0)$; its *transient length* is defined by:

$$\tau(x(0)) = \max\{t \geq 0 : x(t) \text{ enters a cycle for the first time}\}$$

The transient length of the Neural Network is defined as the greatest of such lengths, that is:

$$\tau(A, b) = \max\{\tau(x(0)) : x(0) \in \{0, 1\}^n\}.$$

Obviously $\tau(A, b)$ is finite.

Denote by $\tilde{X}$ the set of all initial conditions which do not belong to a period-2 cycle:

$$\tilde{X} = \{x(0) \in \{0, 1\}^n : x(2) \neq x(0)\}$$

Recall that $\tilde{X}$ is empty iff the transient length of the Neural Network iterated synchronously is null. If $\tilde{X} \neq \phi$ define:

$$e = \min\{-(E_{sy}^*(x(2)) - E_{sy}^*(x(1))) : x(0) \in \tilde{X}\} \tag{3.10}$$

Since $x(0) \in \tilde{X}$ we have $x(2) \neq x(0)$; hence from the proof of Proposition 3.1 the quantity $-(E_{sy}^*(x(2)) - E_{sy}^*(x(1)))$ is strictly positive. Then $\tilde{X} \neq \phi$ implies $\tau(A, b) > 0$. We note $e = 0$ if $\tilde{X} = \phi$.

Proposition 3.4. [G8] Let A be symmetric. Then the transient length $\tau(A, b)$ of the Neural Network iterated synchronously is bounded by:

$$\tau(A, b) \leq \frac{1}{e}(\|A\|_1 + 3\|2b - A\bar{1}\|_1 - 2\sum_{i \in I} e_i) \quad \text{if } e > 0 \tag{3.11}$$

$$\tau(A, b) = 0 \quad \text{if } e = 0$$

Proof. We have $\tau(A, b) = 0$ iff $e = 0$. Assume $e > 0$. Take any trajectory $\{x(t)\}_{t \geq 0}$ with $\tau(x(0)) = \tau(A, b)$ and denote $t_0 = \tau(A, b) + 1$. Then for any $2 \leq t \leq t_0$ we have $x(t) \neq x(t - 2)$. By definition of e we deduce:

$$E_{sy}^*(x(t)) \leq E_{sy}^*(x(t - 1)) - e \quad \text{for any } 1 \leq t \leq \tau(A, b) + 1.$$

Then $E_{sy}^*(x(t_0)) \leq E_{sy}^*(x(1)) - (t_0 - 1)e$. From Proposition 3.1 we find

$$-\|A\|_1 - 2\|2b - A\bar{1}\| \leq E^*(x(t_0)) \leq E^*(x(1)) - \tau(A, b)e$$

$$\leq -2\sum_{i \in I} e_i + \|2b - A\bar{1}\|_1 - \tau(A, b)e$$

Then:

$$\tau(A,b) \le \frac{1}{e}(\|A\|_1 + 3\|2b - A\bar{1}\|_1 - 2\sum_{i \in I} e_i). \quad \blacksquare$$

When the matrix A takes values on the integers, the quantities e_i, e can be controlled and an explicit bound of $\tau(A, b)$ in terms of A, b can be given.

Previously let us make some few remarks. For $a = (a_1, ..., a_n) \in \mathbb{Z}^n$ we have the equality:

$$\mathbb{1}(\sum_{j \in I} a_j u_j - \theta) = \mathbb{1}(\sum_{j \in I} a_j u_j - ([[\theta]] - \frac{1}{2})) \quad \forall u = (u_1, ..., u_n) \in \mathbb{Z}^n \quad (3.12)$$

where $[[\theta]] = \min\{m \in \mathbb{Z} : m \ge \theta\}$. In fact expression (3.12) follows from the equivalence $\sum_{j \in I} a_j u_j \ge \theta \Leftrightarrow \sum_{j \in I} a_j u_j \ge [[\theta]] - \frac{1}{2}$.

Hence if A is an integer matrix and $x(t+1) = \bar{\mathbb{1}}(Ax(t) - b)$ is the synchronous iteration with $x(t) \in \{0,1\}^n$, we can always assume the components b_i of the threshold vector to be of the form $b_i = m_i + \frac{1}{2}$ with $m_i \in \mathbb{Z}$.

Proposition 3.5. [G2] Let A be an integer symmetric matrix. Then the transient length of the network is either null or it is bounded by:

$$\tau(A,b) \le \frac{1}{2}(\|A\|_1 + 3\|2b - A\bar{1}\|_1 - n) \quad (3.13)$$

where $b = (b_i : i \in I)$ is a threshold vector with $b_i = m_i + \frac{1}{2}$, $m_i \in \mathbb{Z}$ for any $i \in I$.

Proof. From the form of A and b we get:

$$e_i = \min\{|\sum_{j \in I} a_{ij} u_j - b_i| : u \in \{0,1\}^n\} \ge \frac{1}{2}$$

Now from the expression

$$\Delta_t E_{sy}^* = 4\Delta_t E_{sy} = -4\sum_{i \in I}((x_i(t) - x_i(t-2))(\sum_{j \in I} a_{ij} x_j(t) - b_i))$$

we deduce: if $e > 0$ then $e \ge 4 \cdot 1 \cdot \frac{1}{2} = 2$. Hence by replacing these bounds on (3.11) we obtain the result. $\blacksquare$

There exist symmetric Neural Networks where the bound (3.13) is attained. For instance take $b_i = \frac{1}{2}$ for any $i \in I$ and $A = (a_{ij} : i, j \in I)$ the following symmetric matrix: $a_{i,i-1} = a_{i,i+1} = 1$ for $i = 2, ..., n-1$, $a_{1,2} = a_{n,n-1} = a_{n,n} = 1$ and $a_{ij} = 0$ for any other pair (i,j).

Hence the synchronous iteration scheme is given by:

$$x_1(t+1) = \mathbb{1}\left(x_2(t) - \frac{1}{2}\right)$$

$$x_i(t+1) = \mathbb{1}\left(x_{i-1}(t) + x_{i+1}(t) - \frac{1}{2}\right) \qquad \text{for } 2 \le i \le n-1$$

$$x_n(t+1) = \mathbb{1}\left(x_{n-1}(t) + x_n(t) - \frac{1}{2}\right)$$

We have $\|A\|_1 = 2n - 1$, $\|2b - A\bar{1}\|_1 = \|(1,1,...,1) - (1,2,...,2)\|_1 = n - 1$, so the upper bound (3.13) is: $\frac{1}{2}(2n - 1 + 3n - 3 - n) = 2(n - 1)$. We shall prove that the trajectory of $x(0) = (1, 0, ..., 0)$ possesses a transient $\tau(x(0)) = 2(n - 1)$.

For simplicity assume n is even. The initial condition $x(0)$ evolves as follows:

$$x(0) = (1, 0, 0, 0, ..., 0, 0)$$
$$x(1) = (0, 1, 0, 0, ..., 0, 0)$$
$$x(2) = (1, 0, 1, 0, ..., 0, 0)$$
$$x(3) = (0, 1, 0, 1, ..., 0, 0)$$

$$\vdots$$

$$x(2i - 1) = (0, 1, ..., 1, 0, 1, 0, ..., 0, 0) \quad \text{for } 2i - 1 < n - 1$$

$$\vdots$$

$$x(n - 1) = (0, 1, 0, 1, ..., 0, 1, 0, 1)$$
$$x(n) = (1, 0, 1, 0, ..., 1, 0, 1, 1)$$

$$\vdots$$

$$x(2n - 3) = (0, 1, 1, 1, ..., 1)$$
$$x(2n - 2) = (1, 1, 1, 1, ..., 1) \qquad \text{which is a fixed point}$$

Hence $\tau(x(0)) = 2(n - 1)$, which is just the bound given by (3.13).

We are also able to obtain bounds for the transient length of Neural Networks defined on regular graphs, i.e. those networks whose neighbour structure is given by a set of non-oriented edges V (so $(i,j) \in V$ iff $(j,i) \in V$) satisfying $|\{j/(i,j) \in V\}| = k$ for any $i \in I$. The number k is called the size of the uniform neighbourhood.

Proposition 3.6. [GFP] Consider a Neural Network of size n with uniform neighbourhood V of size k and interactions given by $a_{ij} = a_{ji} \in \{-1, 1\}$ when $(i,j) \in V$

and $a_{ij} = 0$ otherwise. Then the transient length τ of its synchronous iteration is bounded by:

$$\tau \leq \frac{1}{2}(nk + 9nk - n) \leq \frac{1}{2}n(10k - 1) \leq 5nk \qquad (3.14)$$

Proof. If $|b_i| > k$ the i-th cell remains fixed from the first step so we can consider only those i such that $b_i \in [-k, k]$. Then in order to get an upper bound we can suppose that each $i \in I$ satisfies $b_i \in [-k, k]$. The assumptions of Proposition 3.5 hold. In our case $\|A\|_1 = nk, 3\|2b - A\bar{1}\|_1 \leq 9kn$, hence the bound (3.14) follows from (3.13). ∎

Remark. Clearly, transient lengths, for fixed k, are linear in the size, n, of the network. In practice the convergence is faster than in expression (3.14) but no better theoretical bounds are known.

3.3. Sequential Iteration

As we defined in Chapter 1, the sequential updating is:

$$x_i(t + 1) = \mathbb{1}\Big(\sum_{j<i} a_{ij}x_j(t + 1) + \sum_{j\geq i} a_{ij}x_j(t) - b_i\Big)$$
$$\text{for } i \in I = \{1, ..., n\}, \text{ and } x(0) \in \{0,1\}^n \qquad (3.15)$$

Often it is convenient to write the previous iteration as a procedure whose steps are the following:

(0) Take $x \in \{0,1\}^n$, assign $i \leftarrow 1$

(1) $i \leftarrow i + 1$ (we consider $1 \equiv n + 1$),

(2) update the i-th cells: $x_i \leftarrow \mathbb{1}\Big(\sum_{j\in I} a_{ij}x_j - b_i\Big)$ \qquad (3.16)

(3) if the current vector is not an steady state go to 1

This form of writing the sequential iteration was used in the study of the Hopfield model of associative memories [Ho2] and it is very well adapted for the Lyapunov functional techniques.

Recall that a one-time step of the original evolution (3.15) corresponds to a block of n consecutives iterates of procedure (3.16). To set a common scale, we introduce times $s = t + \frac{i-1}{n}$ for $t \in \mathbb{N}$, $i \in I = \{1, ..., n\}$. We define:

$$x\Big(t + \frac{i-1}{n}\Big) = (x_1(t + 1), ..., x_{i-1}(t + 1), x_i(t), ..., x_n(t)) \qquad \text{for } i \in I \qquad (3.17)$$

where $x_j(t)$ corresponds to the state of site j at time t of sequential iteration.

Then the evolution $x(s) \rightarrow x(s + \frac{1}{n})$ corresponds to one step of procedure (3.16) and the sequential iteration (3.15) is read in the integer values of time s.

In the new set of time $\frac{1}{n} I\!N = \{s = t + \frac{i-1}{n} : t \in I\!N, i \in I\}$ equation (3.17) can be written as a dynamical one. In fact if $s = t + \frac{i-1}{n}$, $t \in I\!N$, $i \in I$, it suffices to put:

$$x_k\left(s + \frac{1}{n}\right) = x_k(s) \qquad \text{if } k \neq i$$

$$x_i\left(s + \frac{1}{n}\right) = 1\!\!1\left(\sum_{j \in I} x_k(s) - b_i\right) \qquad (3.18)$$

which is just another form of writing (3.16). We refer to sequential iteration indistinctly as evolutions (3.15) or (3.18).

Proposition 3.8. [FG2,Ho2] Let A be a real symmetric matrix with $\text{diag} A \geq 0$. Then the functional:

$$E_{seq}(x(s)) = -\frac{1}{2} < x(s), Ax(s) > + < b, x(s) > \qquad (3.19)$$

is decreasing for the sequential iteration of the Neural Network. Furthermore if $x(s + \frac{1}{n}) \neq x(s)$ then $E_{seq}(x(s + \frac{1}{n})) < E_{seq}(x(s))$.

Proof. Suppose $s = t + \frac{i-1}{n}$ with $t \in I\!N$, $i \in I$. Let $\Delta_s E = E(x(s + \frac{1}{n})) - E(x(s))$. By symmetry of A we have:

$$\Delta_s E_{seq} = -\left(x_i\left(s + \frac{1}{n}\right) - x_i(s)\right)\left(\sum_{j \neq i} a_{ij} x_j(s) - b_i\right) - \frac{1}{2} a_{ii}\left(x_i^2\left(s + \frac{1}{n}\right) - x_i^2(s)\right)$$

$$= -\left(x_i\left(s + \frac{1}{n}\right) - x_i(s)\right)\left(\sum_{j \in I} a_{ij} x_j(s) - b_i\right) - \frac{1}{2} a_{ii}\left(x_i\left(s + \frac{1}{n}\right) - x_i(s)\right)^2.$$

Since $a_{ii} \geq 0$ we conclude $\Delta_s E_{seq} \leq 0$, with $\Delta_s E_{seq} < 0$ in the case $x_i(s + \frac{1}{n}) \neq x_i(s)$. ∎

This result clearly implies:

Corollary 3.1. If A is a real symmetric matrix with $\text{diag} A \geq 0$, the cycles of the sequential iteration of the Neural Network are only fixed points. ∎

In the same fashion as done for the synchronous case with (3.6), a formula can be obtained relating the above Lyapunov functional for sequential iteration and the algebraic invariant L_{se}.

Now let us give some bounds of the transient length of the sequential iteration. First recall that if we define:

$$e = \min\{|\sum_{j \in I} a_{ij} u_j - b_i| : i \in I, \ u = (u_1, ..., u_n) \in \{0,1\}^n\} \qquad (3.20)$$

then a lower bound for the decreasing rate of the Lyapunov functional is given by:

$$\text{if } x_i\left(s + \frac{1}{n}\right) \neq x_i(s) \text{ then } |\Delta_s E_{seq}| \geq e + \frac{1}{2} a_{ii} \qquad (3.21)$$

Now let us introduce a more appropriate Lyapunov functional to study the transient length. Define:

$$E_{seq}^*(x(s)) = - < 2x(s) - \bar{1}, A(2x(s) - \bar{1}) > + < b - A\bar{1}, 2x(s) - \bar{1} > \qquad (3.22)$$

Proposition 3.9. Let A be symmetric with $\text{diag} A \geq 0$. The difference $\Delta_s E_{seq}^* = E_{seq}^*(\bar{x}(s + \frac{1}{n})) - E_{seq}^*(\bar{x}(s))$ satisfies $\Delta_s E_{seq}^* = 4\Delta_s E_{seq}$; hence $E_{seq}^*(x(s))$ is an strictly decreasing Lyapunov functional.

Proof. From the symmetry of A we can show that:

$$\Delta_s E_{seq}^* = -\left(2\bar{x}_i\left(s + \frac{1}{n}\right) - 2\bar{x}_i(s)\right)\left(2 \sum_{j \in I} a_{ij} \bar{x}_j(s) - 2b_i\right)$$

$$= -\frac{1}{n} a_{ii} - \left(2\bar{x}_i\left(s + \frac{1}{n}\right) - 2\bar{x}_i(s)\right)^2 = 4\Delta_s E_{seq}. \quad \blacksquare$$

Since the variables $2x_i(s) - 1$ take values on the set $\{-1, 1\}$, then it is direct to show that $|E_{seq}^*(x(s))| \leq \frac{1}{2}\|A\|_1 + \|2b - A\bar{1}\|_1$. On the other hand, $|\Delta_s E_{seq}| \geq e + \frac{1}{2} \min_i a_{ii}$, hence $|\Delta_s E_{seq}^*| \geq 4(e + \frac{1}{2} \min_i a_{ii})$ so we deduce:

Proposition 3.10. If A is symmetric and $\text{diag} A \geq 0$ the transient length of sequential iteration $\tau_{seq}(A, b)$ is bounded by:

$$\tau_{seq}(A, b) \leq \frac{1}{4\left(e + \frac{1}{2} \min a_{ii}\right)}\left(\|A\|_1 + 2\|2b - A\bar{1}\|_1\right) \quad \blacksquare \qquad (3.23)$$

Now we shall obtain another type of bound for τ_{seq} (A, b) by taking into account the eigenvalues of the symmetric matrix of connections. Let us call λ_{min}

(resp. λ_{max}) the minimum (resp. the maximum) of the eigenvalues of A. We recall that for any unitary vector $z \in \mathbb{R}^n$

$$\lambda_{min} \leq < z, Az > \leq \lambda_{max}.$$

Since any vector $y \in \{-1,1\}^n$ posseses a norm $< y, y >^{\frac{1}{2}} = \sqrt{n}$ we have

$$\lambda_{min} n \leq < y, Ay > \leq \lambda_{max} n \quad \forall y \in \{-1,1\}^n \tag{3.24}$$

Since the vectors $2x(s) - 1$ belong to $\{-1,1\}^n$, then we can bound $E^*_{seq}(x(s))$ defined in (3.44) in either of the following two forms:

$$-\lambda_{max} n - \|2b - A\bar{1}\|_1 \leq E^*_{seq}(x(t)) \leq -\lambda_{min} n + \|2b - A\bar{1}\|_1$$
$$-\lambda_{max} n - \|2b - A\bar{1}\|_2 \sqrt{n} \leq E^*_{seq}(x(t)) \leq -\lambda_{min} n + \|2b - A\bar{1}\|_2 \sqrt{n}$$

By taking the same e as in definition (3.41), we obtain:

Proposition 3.11. If A is symmetric and $\text{diag} A \geq 0$ then we have the bounds:

$$\tau_{seq}(A, b) \leq \frac{1}{4e}((\lambda_{max} - \lambda_{min})n + 2\|2b - A\bar{1}\|_1)$$

$$\tau_{seq}(A, b) \leq \frac{1}{4e}((\lambda_{max} - \lambda_{min})n + 2\|2b - A\bar{1}\|_2 \sqrt{n})$$

Corollary 3.2. Assume that A is symmetric, $\text{diag} A \geq 0$, $a_{ij} \in \mathbb{Z}$ for any $i, j \in I$ and $b_i = \frac{1}{2} \sum_{j \in I} a_{ij} \ \forall i \in I$ (the last condition defines a self-dual network [GO2]). Then:

$$\tau_{seq}(A, b) \leq \frac{1}{4}(\lambda_{max} - \lambda_{min})n$$

Proof. In this case $e = \frac{1}{2}$. ∎

Let us briefly discuss the bounds here obtained in the context of the Hopfield model.

In the Hopfield model of Associative Memories [Ho2] a symmetric matrix A is related to p random vectors $S^1, ..., S^p \in \{0,1\}^n$ such that the patterns $\{S^k : k = 1, ..., p\}$ are fixed points of the sequential iteration of the associated neural model i.e.:

$$S^k = \bar{1}(AS^k) \text{ for any } k = 1, ..., p$$

It is further required that for any configuration $x \in \{0,1\}^n$ sufficiently close to S^k in terms of the Hamming distance the sequential dynamics starting from x converges to S^k. In this context the vectors S^k can be interpreted as memorized patterns that may be reconstructed from partial or noise initial information.

In order that the previous requirements are fulfilled, the matrix A is determined by the Hebbs rule, i.e.:

$$a_{ij} = \sum_{k=1}^{p}(2S_i^k - 1)(2S_j^k - 1) \qquad \text{for } i \neq j$$

$$a_{ij} = 0 \qquad \text{if } i = j$$

(3.25)

The threshold vector is taken as 0, but as $a_{ij} \in \mathbb{Z}$ we can assume $b_i = -\frac{1}{2} \ \forall i \in \mathbb{Z}$ without changing the dynamics.

From the definition $\|A\|_1 \leq pn^2$, $\|2b - A\bar{1}\|_1 \leq n^2 p + n$, $e \geq \frac{1}{2}$. As $\text{diag} A = 0$ the result obtained in Proposition 3.10. holds, so the transient of the Hopfield model is bounded by:

$$\tau_H \leq \frac{1}{2}(2pn^2 + n) = 0(2pn^2)$$

Hence the convergence is polynomial.

3.4. Tie Rules for Neural Networks

Tie cases occur when the configurations lie in the hyperplane of discontinuity of the threshold function. We assume that interactions are symmetric, $a_{ij} = a_{ij}$, and $a_{ii} = 0$ so the configuration $x \in \{0,1\}^n$ is in tie case if $\sum_{j \neq i} a_{ij} x_j - b_i = 0$ for some $i \in I$. Several rules have been devised to handle this problem and some of them are relevant for the Ising model [GFP]. Here we examine a particular one, whose sequential updating study must be conducted based on new functionals.

Let I_0, I_1 be a partition of $I = \{1, ..., n\}$. We consider a rule such that in tie case at site i, i.e. if $\sum_{j \neq i} a_{ij} x_j - b_i = 0$, the state x_i is preserved for $i \in I_0$ and reversed for $i \in I_1$. More precisely, we define the following function $F(x) = (F_i(x) : i \in I)$ from $\{0,1\}^n$ into itself,

$$F_i(x) = \begin{cases} \mathbb{1}(\sum_{j \neq i} a_{ij} x_j - b_i) & \text{if } \sum_{j \neq i} a_{ij} x_j - b_i \neq 0 \\ x_i & \text{if } i \in I_0 \text{ and } \sum_{j \neq i} a_{ij} x_j - b_i = 0 \\ 1 - x_i & \text{if } i \in I_1 \text{ and } \sum_{j \neq i} a_{ij} x_j - b_i = 0 \end{cases}$$

(3.26)

To give this function a familiar form we define new diagonal elements $(a_{ii} : i \in I)$ of the matrix A in such a way that they satisfy the following conditions:

$$0 < a_{ii} < -\min\{\sum_{j \neq i} a_{ij} u_j - b_i < 0 : u \in \{0,1\}^n\} \text{ if } i \in I_0$$

$$0 > a_{ii} > -\min\{\sum_{j \neq i} a_{ij} u_j - b_i > 0 : u \in \{0,1\}^n\} \text{ if } i \in I_1$$

$$(3.27)$$

Then the function (3.26) can be written in the standard form $f(x) = \bar{\mathbb{1}}(Ax - b)$.

Since the matrix of connections A is symmetric, the synchronous updating of the tie case $x(t+1) = F(x(t))$ is described by the general results we have obtained, and hence there exists a Lyapunov functional which gives a bound for the transient and there exists only one or two cycles. But the sequential updating of the tie case cannot be studied with the tools we have developed because $\text{diag} A$ contains negative entries $(a_{ii} < 0$ if $i \in I_1)$. In fact the sequential updating of a tie case can possess cycles different from fixed points. In spite of these differences, a relaxed monotonic functional may be used to get partial information about periods of a tie sequential updating.

Recall that the sequential updating of the tie case is written in time set $\frac{1}{n} I\!N = \{s = t + \frac{i-1}{n} : t \in I\!N, i \in I\}$. Its evolutions equation is:

$$x_k\left(s + \frac{1}{n}\right) = x_k(s) \text{ if } k \neq i$$
$$\text{for } s = t + \frac{i-1}{n} \qquad (3.28)$$
$$x_i\left(s + \frac{1}{n}\right) = F_i(x(s))$$

Define:

$$E_{seq}(x(s)) = -\frac{1}{2} \sum_{i \in I} x_i(s) \sum_{j \neq i} a_{ji} x_j(s) + \sum_{i \in I} b_i x_i(s) \qquad (3.29)$$

A calculation similar to those performed in previous sections yields:

$$\Delta_s E_{seq} = E_{seq}(x(s + \frac{1}{n})) - E_{seq}(x(s)) = -(x_i(s + \frac{1}{n}) - x_i(s))(\sum_{j \neq i} a_{ij} x_j(s) - b_i)$$

Hence $x_i(s + \frac{1}{n}) = x_i(s)$ implies $\Delta_s E_{seq} = 0$. If $x_i(s + \frac{1}{n}) \neq x_i(s)$ and $i \in I_0$ we get $\Delta_s E_{seq} < 0$. If $x_i(s + \frac{1}{n}) \neq x_i(s)$ and $i \in I_1$ we get $\Delta_s E_{seq} = 0$ iff $\sum_{j \neq i} a_{ij} x_j(s) - b_i = 0$, $\Delta_s E_{seq} < 0$ otherwise. Then we obtain:

Proposition 3.13. [GFP] $E_{seq}(x(s))$ is a Lyapunov functional for sequential updating of a tie case. Now let $s = t + \frac{i-1}{n}$ with $t \in \mathbb{N}$, $i \in I$:

$$\text{for } i \in I_0 \text{ we have } \Delta_s E_{seq} < 0 \qquad \text{if } x_i\left(s + \frac{1}{n}\right) \neq x_i(s),$$

$$\Delta_s E_{seq} = 0 \qquad \text{if } x_i\left(s + \frac{1}{n}\right) = x_i(s)$$

$$\text{for } i \in I_1 \text{ we have } \Delta_s E_{seq} < 0 \qquad \text{if } x_i\left(s + \frac{1}{n}\right) \neq x_i(s) \text{ and}$$

$$\sum_{j \neq i} a_{ij} x_j(s) - b_i \neq 0$$

$$\Delta_s E_{seq} < 0 \qquad \text{if } x_i\left(s + \frac{1}{n}\right) = x_i(s) \text{ or}$$

$$\sum_{j \neq i} a_{ij} x_j(s) - b_i = 0 \qquad \blacksquare$$

Clearly when $\frac{|I_1|}{|I_0|} \ll 1$ the behaviour of the system approaches that of a frozen system, i.e. only fixed points exist. When $\frac{|I_1|}{|I_0|}$ is sufficiently large the system may have large period when the configurations are in the tie case, but for the particular case of a two-dimensional lattice the tie case is a very strong combinatorial restriction and only small period cycles can appear (see Figure 3.3 for the case $I_1 = I$).

Let us examine some special cases. Take $I = \{1, ..., n\}$. Consider the partition $I_0 = \phi$, $I_1 = I$.

Take the threshold $b = \bar{1}$ so the functions F_i of (3.26) can be written as:

$$F_i(x_{i-1}, x_i, x_{i+1}) = \begin{cases} 0 & \text{if } x_{i-1} + x_{i+1} - 1 < 0 \\ 1 - x_i & \text{if } x_{i-1} + x_{i+1} - 1 = 0 \\ 1 & \text{otherwise} \end{cases} \qquad (3.30)$$

The sequential order to iterate is the canonical one: $1 < 2 \cdots < n$.

It is easy to see that all configurations are gliders (vehicles) of period n, i.e. they are not bounded when the size of the network varies. Let us iterate the particular configuration (000111100) in a 9-site torus. Its dynamics is shown in Figure 3.1.

$$\begin{matrix}
0 & 0 & 0 & 1 & 1 & 1 & 1 & 0 & 0 \\
0 & 0 & 1 & 1 & 1 & 1 & 0 & 0 & 0 \\
0 & 1 & 1 & 1 & 1 & 0 & 0 & 0 & 0 \\
1 & 1 & 1 & 1 & 0 & 0 & 0 & 0 & 1 \\
1 & 1 & 1 & 0 & 0 & 0 & 0 & 1 & 1 \\
1 & 1 & 0 & 0 & 0 & 0 & 1 & 1 & 1 \\
1 & 0 & 0 & 0 & 0 & 1 & 1 & 1 & 1 \\
0 & 0 & 0 & 0 & 1 & 1 & 1 & 1 & 0 \\
0 & 0 & 0 & 1 & 1 & 1 & 1 & 0 & 0
\end{matrix}$$

Figure 3.1. 8-cycle for sequential iteration of scheme (3.30) for $n = 9$.

For the same case pick $I_0 \neq \phi$, for instance $I_0 = \{3\}$. Then,

$$F_3\left(x_2, x_3, x_4\right) = \begin{cases} 0 & \text{if } x_2 + x_4 - 1 < 0 \\ x_3 & \text{if } x_2 + x_4 - 1 = 0 \\ 1 & \text{otherwise} \end{cases}$$

It can be shown that cell 3 freezes the dynamics, i.e. any initial configuration converges to a fixed point, as we show in Figure 3.2.

$$
\begin{array}{c}
3 \\
\downarrow
\end{array}
\qquad\qquad
\begin{array}{c}
3 \\
\downarrow
\end{array}
$$

$$\begin{matrix}
0 & 0 & 0 & 0 & 0 & 0 & 1 & 0 & 0 & \quad & 0 & 0 & 0 & 0 & 1 & 1 & 1 & 1 & 0 \\
0 & 0 & 0 & 0 & 0 & 1 & 0 & 0 & 0 & \quad & 0 & 0 & 0 & 1 & 1 & 1 & 1 & 0 & 0 \\
0 & 0 & 0 & 0 & 1 & 0 & 0 & 0 & 0 & \quad & 0 & 0 & 0 & 0 & 0 & 0 & 0 & 0 & 0 \\
0 & 0 & 0 & 0 & 0 & 0 & 0 & 0 & 0 & \quad & 0 & 0 & 0 & 0 & 0 & 0 & 0 & 0 & 0
\end{matrix}$$

Figure 3.2. Freezing dynamics for 1-dimensional sequential tie iteration. The third cell belongs to I_0.

Other examples for the $1 - D$ torus and the neighbourbood $V_0 = \{-2, -1, 0, 1, 2\}$ were given in Figure 1.6.

The behaviour of sequential iteration tie-rules is more difficult to analyze in higher dimensions. For instance, consider the two dimensional torus $\mathbb{Z}_n^2$ and let $\{I_0, I_1\}$ be a partition of its sites. Define the local rules in the von-Neumann neighbourhood $V_0^N = \{(-1,0), (1,0), (0,0), (0,-1), (0,1)\}$ as follows.

For $i = (i_1, i_2)$, $j = (j, j') \in \mathbb{Z}_n^2$

$$F_i\left(x_j : j \in V_0^N + i\right) = \begin{cases} \mathbb{1}\left(\displaystyle\sum_{j \in V_0^N + i} x_j - 2\right) \text{ and } \displaystyle\sum_{j \in V_0^N + i} x_j - 2 \neq 0 \\[2ex] x_i & \text{if } i \in I_0 \text{ and } \displaystyle\sum_{j \in V_0^N + i} x_j - 2 = 0 \\[2ex] x_i - 1 & \text{if } i \in I_1 \text{ and } \displaystyle\sum_{j \in V_0^N + i} x_j - 2 = 0 \end{cases}$$

For $n = 30$ we simulate the sequential iteration from left to right and from top to bottom. The simulation procedure depends on the density, $d = |I_1|/|\mathbb{Z}_n^2| = |I_1|/n^2$. Given d, we choose at random 10 initial conditions and observe the number of different configurations, T, obtained before getting a cycle. When this number is $\geq 10,000$ we stop the dynamics and we take $T = 10,000$. For a fixed density we plot $< T >$ the average among the dynamics for the 10 initial conditions. The graph is exhibited in Figure 3.3; it was built by Matamala. Clearly for small densities, $d < 0.40$, small periods appears and the iteration converges to a frozen steady state. But for $d > 0.40$ there appears a change in behaviour: large periods and/or transient behaviours. In particular for $d \approx 0.75$, $< T > \geq 10,000$. When d approaches 1 (i.e. $|I_1| = |\mathbb{Z}_n^2|$), $< T >$ diminishes as we predicted in the remarks of Proposition 3.13. More precisely, for $d \approx 1$ the periods are small, $< T > \approx 6$ and we obtain a new frozen situation.

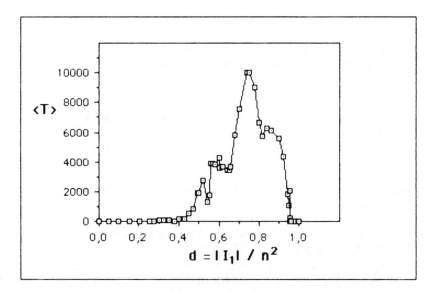

Figure 3.3. Average cycles for the sequential dynamics of the two-dimensional 30×30 torus. $< T >$ is the average of the number of differents configuration among ten random initial conditions. d is the density of sites belonging to I_1.

Typical patterns, for different values of d are exhibited in Figure 3.4.

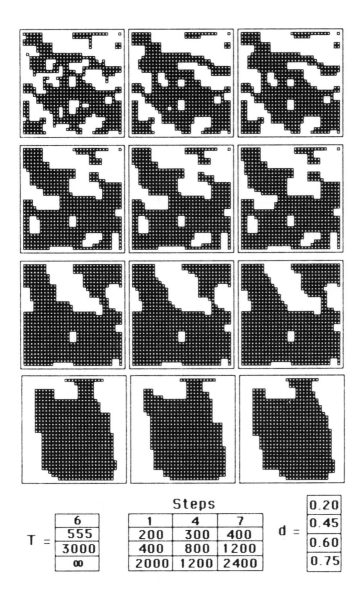

Figure 3.4. Sequential dynamics of tie rules in a two-dimensional 30×30 torus.

3.5. Antisymmetrical Neural Networks

Let us consider the following synchronous iteration:

$$x_i(t+1) = 1\!\!1 \left(\sum_{j \in I} a_{ij} x_j(t) - \frac{1}{2} \sum_{j \in I} a_{ij} \right) \text{ for } i \in I \tag{3.31}$$

or equivalentely $x_i(t+1) = 1\!\!1 \left(\sum_{j \in I} a_{ij} (2x_j(t) - 1) \right)$.

The matrix of interactions A is supposed to be antisymmetric i.e.:

$$a_{ij} = -a_{ji} \text{ for any } i, j \in I \tag{3.32}$$

in particular this implies $a_{ii} = 0$ for any $i \in I$.

We shall also suppose:

$$\sum_{j \in I} a_{ij} u_j - \frac{1}{2} \sum_{j \in I} a_{ij} \neq 0 \text{ for any } u = (u_1, ..., u_n) \in \{0,1\}^n \tag{3.33}$$

Recall that for some antisymmetric matrices A (for instance $A = \begin{pmatrix} 0 & 1 \\ -1 & 0 \end{pmatrix}$) it is not possible to obtain another antisymmetric matrix $\tilde{A}$ verifying condition (3.33) in such a way that the evolution of iteration (3.31) is the same for both A and $\tilde{A}$.

Define the functional:

$$E_{sy}(x(t)) = - \sum_{i \in I} \left((2x_i(t) - 1) \sum_{j \in I} a_{ij} (2x_j(t-1) - 1) \right) \tag{3.34}$$

Consider $\Delta_t E_{sy} = E_{sy}(x(t)) - E_{sy}(x(t-1))$. From the antisymmetry of A we get:

$$\Delta_t E_{sy} = - \sum_{i \in I} \left(((2x_i(t) - 1) + (2x_i(t-2) - 1)) \left(\sum_{j \in I} a_{ij} (2x_j(t-1) - 1) \right) \right)$$

$$= -4 \sum_{i \in I} \left((x_i(t) + (x_i(t-2) - 1) \left(\sum_{j \in I} a_{ij} x_j(t-1) - \frac{1}{2} \sum_{j \in I} a_{ij} \right) \right)$$

Take the i-th component of the above sum:

$$(\Delta_t E_{sy})_i = -2(x_i(t) + x_i(t-2) - 1) \left(\sum_{j \in I} a_{ij} (2x_j(t-1) - 1) \right)$$

Obvioulsy if $x_i(t) \neq x_i(t-2)$ the first factor is null so $(\Delta_t E_{sy})_i = 0$.
If $x_i(t) = x_i(t-2) = 1$ then $(\Delta_t E_{sy})_i = -2(\sum_{j \in I} a_{ij}(2x_j(t-1)-1)) < 0$. When
$x_i(t) = x_i(t-2) = 0$ we also have $(\Delta_t E_{sy})_i = 2(\sum_{j \in I} a_{ij}(2x_j(t-1)-1)) < 0$. Hence:

Proposition 3.14. [G10] $E_{sy}(x(t))$ is a Lyapunov functional, and it is strictly
decreasing iff $x(t) = x(t-2)$. Hence the orbits have a period $T = 4$.

Proof. If $\{x(t) : t \in \mathbb{Z}_T\}$ is an orbit of period T of the synchronous iteration
then $E_{sy}(x(t)) = E_0$ is constant for any t. Hence $x_i(t+2) \neq x_i(t)$ for all $i \in I$.
This necessarily implies $x(t+4) = x(t)$ for all t. ∎

Proposition 3.15. [G10] Assume the antisymmetric matrix A is an integer matrix
satisfying condition (3.33). Let $\tau_{as}(A)$ be the transient length of the iteration (3.31)
an antisymmetric network. Then:

$$\tau_{as}(A) \leq \frac{1}{2}\|A\|_1 \qquad (3.35)$$

Proof. As $2x_i(t) - 1 \in \{-1,1\}$ we have $E_{sy}(x(t)) \geq -\|A\|_1$. On the other hand
if $(\Delta_t E_{sy})_i < 0$ we have $|(\Delta_t E_{sy})_i| = 2|\sum_{j \in I} a_{ij}(2x_j(t-1)-1| \geq 2$ because A is an
integer matrix. By antisymmetry of A the number $\sum_{j \in I} a_{ij}(2u_j - 1)$ is odd for any
$u = (u_1, ..., u_n) \in \{0,1\}^n$.
  As in Proposition 3.4, we can show that the bound (3.35) holds. ∎

For the sequential updating:

$$x_i(t+1) = \mathbb{1}(\sum_{j=1}^{i-1} x_{ij}x_j(t+1) + \sum_{j=i}^{n} a_{ij}x_j(t) - \frac{1}{2}\sum_{j \in I} a_{ij}) \qquad (3.36)$$

similar results are obtained. In fact, we define:

$$E_{seq}(x(t)) = -\sum_{i \in I}(2x_i(t) - 1)(\sum_{j \in I} a_{ij}(2x_j(t) - 1))$$

Let us update the i-th component; we get:

$$\Delta_t E_{seq} = -(2x_i(t+1) - 1 + 2x_i(t) - 1)(\sum_{j \in I} a_{ij}(2x_j(t) - 1))$$

$$= -4(x_i(t+1) + x_i(t) - 1)(\sum_{j \in I} a_{ij}x_j(t) - \frac{1}{2}\sum_{j \in I} a_{ij}).$$

Hence $\Delta_t E_{seq} = 0$ iff $x_k(t+1) \neq x_k(t)$ and $(\Delta_t E_{seq}) < 0$ otherwise.

As in Proposition 3.14, we can show that the sequential updating of ansymmetric matrices admits only cycles of length 2. When A is an integer matrix we can also bound its transient $\tau(A)$ by $\tau(A) \leq \frac{1}{2}\|A\|_1$.

Remarks.

1. It is interesting to point out that in the symmetric case we found two-cycle and fixed point behaviour for the synchronous and for sequential iterations respectively, while in the antisymmetric case we similarly obtained a ratio 2:1 for synchronous to sequential 4- and 2-cycles respectively.

2. Hypothesis (3.33) is necessary for the above periodic behavior. For instance, consider the $n \times n$ two-dimensional torus and the local rule

$$\vec{\forall i} = (i_1, i_2) \in \{1, ..., n\} \times \{1, ..., n\} :$$
$$x_{\vec{i}}(t+1) = \mathbb{1}\big(x_{(i_1-1,i_2)}(t) + x_{(i_1,i_2+1)}(t) - x_{(i_1+1,i_2)}(t) - x_{(i_1,i_2-1)}(t)\big)$$

Clearly the matrix A associated to this network is antisymmetric but hypothesis (3.33) does not hold.

It is not difficult to verify that the configuration

$$x(0) = \begin{pmatrix} 0 & 1 & \cdots & 1 \\ 0 & 1 & \cdots & 1 \\ & \vdots & & \\ 0 & 1 & \cdots & 1 \end{pmatrix} \in \{0,1\}^{n^2}$$

belongs to an n-cycle, i.e. the column of 0's travels through the n-columns of the torus.

In general, hypotheses (3.33) does not hold in regular arrays. Other examples of antisymmetric synchronous dynamics are given in Figure 4.20.

3.6. A Class of Symmetric Networks with Exponential Transient Length for Synchronous Iteration

The bounds obtained for transient lengths of symmetric matrices depend on $\|A\|_1$ and $\|A\bar{1} - b\|_1$, hence in spite of the symmetry of the network one can expect some classes of symmetrical networks with large transient times to exist. In this paragraph we shall construct one of those networks in a recurrent way.

First we study the synchronous case. To illustrate the idea of our construction consider $n = 2$, $A = \begin{pmatrix} -1 & 1 \\ 1 & 1 \end{pmatrix}$, $b = \begin{pmatrix} -\frac{1}{2} \\ 1 \end{pmatrix}$. This Neural Network is shown in Figure 3.5.

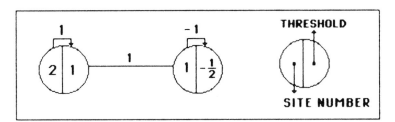

Figure 3.5. Two-site Neural Network.

The dynamics of the point $x(0) = (0,0)$ is: $(0,0) \rightarrow (1,0) \rightarrow (0,1) \rightarrow (1,1)$ the last one being a fixed point. This trajectory travels through all the vectors of the 2-hypercube before reaching the fixed point $(1,1)$, here $\tau(A,b) = 3$. By adding three new sites to the previous network, all the vectors of the 3-hypercube can be traversed. It suffices to take the pair (A,b) as:

$$A = \begin{bmatrix} -1 & 1 & 1 & -2 & 1 \\ 1 & 1 & 1 & -3 & 2 \\ 1 & 1 & 0 & 6 & 0 \\ -2 & -3 & 6 & 0 & 4 \\ 1 & 2 & 0 & 4 & 0 \end{bmatrix} \qquad b = \begin{pmatrix} -1/2 \\ 1 \\ 2 \\ 1 \\ 4 \end{pmatrix}$$

The network is exhibited in Figure 3.6.

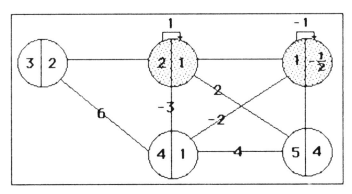

Figure 3.6. Five-site Neural Network. The dotted sites correspond to the network in Figure 3.5.

Recall that the new A, b for $n = 5$ are extensions of A, b defined for $n = 2$.

Now the dynamics of $x(0) = (0,0,0,0,0)$ is:

t	$x_1(t)$	$x_2(t)$	$x_3(t)$	$x_4(t)$	$x_5(t)$	
0	0	0	0	0	0	
1	1	0	0	0	0	
2	0	1	0	0	0	
3	1	1	0	0	0	
4	1	1	1	0	0	
5	1	1	1	1	0	
6	0	0	1	1	1	
7	1	0	1	1	1	
8	0	1	1	1	1	
9	1	1	1	1	1	fixed point

Thus we have traveled through the 3-hypercube, as we consider the states on the sites $\{3, 2, 1\}$. The new sites $\{4, 5\}$ are used as control units to repeat the dynamics of the 2-hypercube twice. In this process, site 4 is on once the first traverse on the 2-hypercube is completed, and at this moment the sites $\{1, 2\}$ switch off. On the other hand, site 5 rules out all the external weights to sites 1 and 2. As a result, these sites repeat the initial trajectory on the 2-hypercube. Now we shall make the above construction more rigorous to obtain exponential transient classes for any length n of the network.

Proposition 3.7. [GM6,GO2] For any n there exists a symmetric Neural Network (A, b) such that its synchronous transient length satisfies $\tau(A, b) \geq 2^{\frac{n}{3}}$.

Proof. Recall that it suffices to show for any n of the form $n = 3m + 2$ that we can construct a symmetric network satisfying $\tau(A, b) \geq 3(2^{m+1} - 1)$. In fact, if $n = 3m + 4$ (or $n = 3m + 3$) we can bound its transient by the transient of the case $3m + 2$. Since $m = \frac{n-4}{3}$ we deduce $\tau(A, b) \geq 3(2^{\frac{n-4}{3}+1} - 1)$ which is $\geq 2^{\frac{n}{3}}$ for $n \geq 10$. For $n < 10$, it follows directly from the network we shall exhibit that $\tau(A, b) \geq 2^{\frac{n}{3}}$.

Hence assume $n = 3m + 2$ with $m \geq 0$. The symmetric Neural Network we shall construct on the set of nodes $I^{(m)} = \{1, ..., 3m + 2\}$ will contain a trajectory as least as large as the following one: $(0, 0, ..., 0) \in \{0, 1\}^{m+2} \rightarrow (0, 1, ..., 0) \in \{0, 1\}^{m+2} \rightarrow (0, 0, 1, ..., 0) \in \{0, 1\}^{m+2} \rightarrow ... \rightarrow (1, 1, 1, ..., 1) \in \{0, 1\}^{m-2}$, which contains 2^{m+2} different points. The other $2m$ sites are used to control the network, i.e. their connections will make possible that such an evolution can be realized.

The construction of the network will be made recursively; at each step we add three nodes, one of them allowing an increase in the length of the above orbit, the other two being used for control. First take $k = 0$, $I^{(0)} = \{1, 2\}$; we shall construct

$A^{(0)}$ (a 2×2 symmetric matrix) and $b^{(0)}$ (a 2-vector). Then we suppose we have constructed a symmetric matrix $A^{(k)}$ and a vector $b^{(k)}$ for $I^{(k)} = \{1, ..., 3k+2\}$ and we give an algorithm to construct a symmetric matrix $A^{(k+1)}$ and a vector $b^{(k+1)}$ on $I^{(k+1)} = \{1, ..., 3(k+1) + 2\}$. The sequence $(A^{(k)}, b^{(k)})$ defined by the algorithm will satisfy:

(i) $A^{(k+1)}$ restricted to $I^{(k)} \times I^{(k)}$ is equal to $A^{(k)}$.

(ii) $b^{(k+1)}$ restricted to $I^{(k)}$ coincides with $b^{(k)}$.

(iii) The initial condition $x(0) = (0, ..., 0) \in \{0, 1\}^{3k+2}$ posseses a transient length $\tau_k = 2^{k+2} + 2^{k+1} - 2$ when we make it evolve synchronously with matrix $A^{(k)}$ and vector $b^{(k)}$.

Then when we put $A = A^{(m)}$, $b = b^{(m)}$ the initial configuration $x(0) = (0, ...0) \in \{0, 1\}^m$ will posseses a transient length $\tau_m = 3(2^{m+1} - 1) = 3(2^{\frac{n-2}{3}+1} - 1)$. Hence it fulfills the properties we have asserted. So take $I^{(0)} = \{1, 2\}$. We construct $A^{(0)} = (a_{ij} : i, j \in I^{(0)})$, $b^{(0)} = (b_1, b_2)$ so as to have the following dynamics for $x(0) = (0, 0)$:

t	$x_1(t)$	$x_2(t)$	
0	0	0	transient behaviour
1	1	0	"
2	0	1	"
$\tau_0 = 3(2^1 - 1) = 3$	1	1	fixed point
5	1	1	"

$$(3.37)$$

It is easy to see that when we improve conditions, say

$$a_{11} < b_1 < 0 < b_2 < \inf(a_{12}, a_{22}), \quad b_1 < a_{11} + a_{12}, a_{21} = a_{12} \qquad (3.38)$$

we get the above dynamics of $x(0)$. Then the transient length of initial condition $x(0)$ is $\tau_0 = 3$.

Now call $C^{(0)}$ the matrix which contains the transient evolution of $x(0) = (0, 0)$ and the first time it attains the fixed point:

$$C^{(0)} = \begin{bmatrix} 0 & 0 \\ 1 & 0 \\ 0 & 1 \\ 1 & 1 \end{bmatrix} \qquad (3.39)$$

Now suppose we have constructed $A^{(k)} = (a_{ij} : i, j \in I^{(k)})$, $b^{(k)} = (b_i : i \in I^{(k)})$ in such a way that $x(0) = (0, ..., 0) \in \{0, 1\}^{3k+2}$ evolves as follows:

t	$x_1(t)$	$\cdots$	$x_{3k+3}(t)$
0			
$\vdots$		$C^{(k)}$	
$\tau_k = 3(2^{k+1} - 1)$			
$\tau_k + 1$	11	$\cdots$	1

$$(3.40)$$

with transient length $\tau_k = 3(2^{k+1} - 1)$.

Now we shall add three elements: $3k+3$, $3k+4$, $3k+5$. We shall make explicit the restrictions of the symmetric matrix of connections $A^{(k+1)}$ and the threshold vector $b^{(k+1)}$ for the initial configuration $x(0) = (0, ..., 0) \in \{0,1\}^{3(k+1)+2}$ to evolve in the following way:

t	$x_1(t)$	$\cdots$	$x_{3k+2}(t)$	$x_{3k+3}(t)$	$x_{3k+4}(t)$	$x_{3k+5}(t)$
0				0	0	0
$\vdots$		$C^{(k)}$		$\cdots$	$\cdots$	$\cdots$
$\tau_k = 3(2^{k+1} - 1)$				0	0	0
$\tau_k + 1$	1	$\cdots$	1	1	0	0
$\tau_k + 2$	1	$\cdots$	1	1	1	1
				1	1	1
$\vdots$		$C^{(k)}$		$\vdots$	$\vdots$	$\vdots$
$\tau_{k+1} = 3(2^{k+2} - 1)$				1	1	1
$\tau_{k+1} + 1$	1	$\cdots$	1	1	1	1

$$(3.41)$$

In the above process we deduce $\tau_k = 3(2^{k+1} - 1)$ because it is the solution to equation $\tau_{k+1} = 2\tau_k + 2 + 1$ with initial condition $\tau_0 = 3$. Also from the recurrence construction it follows that the row $\tau_{k+1} = 3(2^{k+2} - 1)$ is equal to the row $\tau_{k+1} = \tau_{k+1} = 3(2^{k+2} - 1)$, which is the transient length of $x(0)$. We take $C^{(k+1)}$ equal to the matrix formed from row $t = 1$ till row $t = \tau_{k+1}$ and we continue the process:

$$
C^{(k+1)} = \begin{array}{c}
\begin{array}{cccc}
 & & & 0 \; 0 \; 0 \\
C^{(k)} & & & \vdots \; \vdots \; \vdots \\
 & & & 0 \; 0 \; 0 \\
1 & \cdots & 1 \; 1 \; 0 \; 0 \\
1 & \cdots & 1 \; 1 \; 1 \; 1 \\
 & C^{(k)} & \vdots \; \vdots \; \vdots \\
 & & 1 \; 1 \; 1
\end{array}
\quad
\begin{array}{c}
1 \\
\vdots \\
\tau_k \\
\tau_k + 1 \\
\tau_k + 2 \\
\vdots \\
\tau_{k+1}
\end{array}
\end{array}
\qquad (3.42)
$$

Hence we must give explicit constraints on $A^{(k+1)}$, $b^{(k+1)}$ for the above evolution of $x(0)$ to be possible and we must also show that this system of constraints admits a solution.

First, the coefficients a_{ij} of $A^{(k+1)}$ for $i, j \in I^{(k)}$ are the same as those for $A^{(k)}$, and the coefficients b_i of $b^{(k+1)}$ for $i \in I^{(k)}$ are the same as those of $b^{(k)}$.

For $x_i(\tau_k + 1) = 1$ for each $i \in I^{(k)}$, we must have:

$$\gamma_i^{(k)} = \sum_{j \in I^{(k)}} a_{ij} > b_i \text{ for each } i \in I^{(k)} \qquad (3.43)$$

Recall that (3.43) holds for $k = 0$ (see (3.38)); we suppose it holds, by recurrence hypothesis, for k.

On the other hand the condition $x_{3k+3}(t) = 0$ for $t = 0, ..., \tau_k$ is implied by the stronger condition:

$$b_{3k+3} > \kappa_{3k+3} \text{ where } \kappa_{3k+3} = \sup_{L' \subset I^{(k)}, L' \neq I^{(k)}} \sum_{j \in L'} a_{3k+3,j} \qquad (3.44)$$

Equality $x_{3k+3}(\tau_k + 1) = 1$ is implied by the condition:

$$\sum_{j \in I^{(k)}} a_{3k+3,j} > b_{3k+3} \qquad (3.45)$$

Besides, $x_i(\tau_k + 2) = 1$ for each $i \in I^{(k)}$ follows from the inequality:

$$\sum_{j \in I^{(k)}} a_{ij} + a_{i,3k+3} > b_i \quad \forall i \in I^{(k)} \qquad (3.46)$$

In order to verify conditions (3.44), (3.45), (3.46) we take: $a_{i,3k+3} = a_{3k+3,i} > 0$, so $\kappa_{3k+3} < \sum_{j \in I^{(k)}} a_{3k+3,j}$, so we choose b_{3k+3} satisfying (3.44), (3.45) which do always exist. Inequality (3.46) follows from the positiveness of $a_{i,3n+3}$ and from (3.43).

To get $x_{3k+3}(\tau_k + t + 1) = 1$ for $t = 1, 2, ..., (\tau_{k+1} - \tau_k)$, we improve

$$a_{3k+3,3k+3} > \sup_{L' \subset I^{(k)}} \left(-\left(\sum_{j \in L'} a_{3k+3,j} + a_{3k+3,3k+4} + a_{3k+3,3k+5} \right) + b_{3k+3} \right) \qquad (3.47)$$

for any choice we make of $a_{3k+3,3k+4}$, $a_{3k+3,3k+5}$ which will not depend on $a_{3k+3,3k+3}$.

To get the above dynamics of node $3k+4$ i.e. $x_{3k+4}(t) = 0$ for $t = 0, ..., \tau_k + 1$; $x_{3k+4}(\tau_k + 2) = 1$ and $x_{3k+4}(\tau_k + t + 2) = 1$ for $t = 1, 2, ..., (\tau_{k+1} - \tau_k - 1)$ we require that the following inequalities hold:

$$b_{3k+4} > \kappa_{3k+4} \text{ where } \kappa_{3k+4} = \sup_{L' \subset I^{(k)}} \sum_{j \in L'} a_{3k+4,j} \qquad (3.48)$$

$$a_{3k+4,3k+3} = a_{3k+3,3k+4} > b_{3k+4} - \sum_{j \in I^{(k)}} a_{3k+4,j} \qquad (3.49)$$

$$a_{3k+4,3k+4} > \sup_{L' \subset I^{(k)}} \left(-\left\{ \sum_{j \in L'} a_{3k+4,j} + a_{3k+4,3k+3} + a_{3k+4,3k+5} \right\} + b_{3k+4} \right) \qquad (3.50)$$

The conditions on coefficients $a_{3k+4,i} = a_{i,3k+4}$ come from the equality $x_i(\tau_k + 3) = 0$ for any $i \in I^{(k)}$. Then we also need the condition: $\sum_{j \in I^{(k)}} a_{ij} + a_{i,3k+3} + a_{i,3k+4} < b_i$. Hence we impose:

$$a_{3k+4,i} = a_{i,3k+4} < -(\gamma_i^{(k)} + a_{i,3k+3}) \text{ for } i \in I^{(k)} \tag{3.51}$$

where the $\gamma_i^{(k)}$ was defined in (3.43). So $a_{3k+4,i}$ is strictly negative, $a_{3k+4,3k+3}$ is strictly positive. There always exists a solution for the above requirements (3.43)-(3.46).

Analogously, to get $x_{3k+5}(t) = 0$ for $t = 0, ..., \tau_k + 2$, $x_{3k+5}(\tau_k + 3) = 1$, $x_{3k+5}(\tau_k + t + 3) = 1$ for $t = 1, 2, ..., (\tau_{k+1} - \tau_k - 2)$, we impose:

$$b_{3k+5} > \kappa_{3k+5} \text{ where } \kappa_{3k+5} = \sum_{L' \subset I^{(k)} \cup \{3k+3\}} \sum_{j \in L'} a_{3k+5,j} \tag{3.52}$$

$$a_{3k+5,3k+4} = a_{3k+4,3k+5} > b_{3k+5} - \sum_{j \in I^{(k)} \cup \{3k+3\}} a_{3k+5,j} \tag{3.53}$$

$$a_{3k+5,3k+5} > \sup_{L' \subset I^{(k)}} \left(-\left(\sum_{j \in L'} a_{3k+5,j} + a_{3k+5,3k+3} + a_{3k+5,3k+4}\right) + b_{3k+5}\right) \tag{3.54}$$

The last evolution equations, which will only involve conditions on

$$a_{3k+5,j} = a_{j,3k+5} \qquad \text{for } j \in I^{(k)},$$

are the following:

$$x_i(\tau_k + 3 + t) = x_i(t) \qquad \text{for } i \in I^{(k)}, \ t = 0, ..., \tau_k \tag{3.55}$$

In order to satisfy (3.55), let us make the following choice of $a_{3k+5,j}$:

$$a_{3k+5,j} = a_{j,3k+5} = -(a_{3k+4,j} + a_{3k+3,j}) \text{ for } j \in I^{(k)} \tag{3.56}$$

Note that (3.56) implies the equalities:

$$\sum_{j \in L' \cup \{3k+3,3k+4,3k+5\}} a_{ij} = \sum_{j \in L'} a_{ij} \quad \begin{array}{l} \forall \, i \in I^{(k)} \\ \forall \, L' \subset I^{(k)} \end{array} \tag{3.57}$$

We claim that property (3.55) follows from expressions (3.51), (3.46). This is shown by recurrence on $t \geq 0$. In fact for $t = 0$ the equality (3.55) is implied by condition (3.51) and we apply recurrence on $t \geq 1$ by using condition (3.57).

As there always exists a solution for (3.52), (3.53), (3.54), (3.56), we have proved that the evolution of the initial condition $x(0) = (0,...,0) \in \{0,1\}^{3(k+1)+2}$ is the one stated in (3.41). Finally we must remark that condition (3.57) together with (3.47), (3.50), (3.54) imply: $\gamma_i^{(k+1)} = \sum\limits_{j \in I^{(k+1)}} a_{ij} > b_i$ for any $i \in I^{(k+1)}$, then (3.43) can be assumed by recurrence hypothesis. Hence the result. ∎

Remarks.

1. In the foregoing construction we had to add three cells in order to respect the symmetry assumption in A. Otherwise, i.e. if symmetry is not a contraint, it is not dificult to build larger trajectories.

2. Some entries of matrix A in the above proposition grow very fast and also differences between entries may be very large. Hence the data (A, b) for large n may be exponential.

3.7. Exponential Transient Classes for Sequential Iteration

Now we shall prove that the sequential iteration may have, as in the synchronous case, exponential transient lengths. Firstly we shall introduce a construction performed by Tchuente in a more general framework [T2] which allows to simulate any synchronous Neural Network of n sites by a sequential Neural Network of $2n$ sites. Let $I = \{1,...,n\}$, $A = (a_{ij} : i, i \in I)$ be a connection matrix and $b = (b_i : i \in I)$ a threshold vector. We shall iterate (A, b) synchronously. On the set of sites $I = \{1,..., 2n\}$ define $A' = (a'_{ij} : i, j \in I')$ by:

$$a'_{ij} = \begin{cases} 0 & \text{if } |i - j| \leq n \\ a_{i,j-n} & \text{if } 1 \leq i \leq n, \; n+1 \leq j \leq 2n \\ a_{i-n,j} & \text{if } n+1 \leq i \leq 2n, \; 1 \leq j \leq n \end{cases}$$

Clearly A' is symmetric. Take $b' = \binom{b}{b}$. The network defined by (A', b') is updated in a sequential way following the standard order $1 < ...n < n+1 < ... < 2n$. Pick an initial condition $x(0) \in \{0,1\}^n$ and associate to it $x'(0) = (x(0), x(0)) \in \{0,1\}^{2n}$. It is easy to see that if $x(t)$ is the synchronous iterate of $x(0)$ at time t then $x'(t) = (x(t), x(t))$ is the sequential iterate of $x'(0)$ at time t. Consider the pair of symmetric matrix and threshold vector (A, b) for which we proved that the synchronous transient $\tau_{sy}(A, b)$ was exponential, $\tau_{sy}(A, b) \geq 2^{\frac{n}{3}}$. Then the associated pair (A', b') on a network of size $p = 2n$ is such that its sequential transient $\tau_{se}(A, b) \geq 2^{\frac{n}{3}} = 2^{\frac{p}{6}}$. So we may conclude:

Proposition 3.12. [GM6] For any n there exists a symmetric Neural Network (A, b) such that its sequential transient length satisfies $\tau_{seq}(A, b) \geq 2^{\frac{n}{6}}$. ∎

Another approach to exponential transient behaviour of Neural Networks was developed in [HaL], but in their construction they use global information to choose the site to update, which is not a local rule. They compute in each step the quantity $|\sum_j a_{ij} x_j - b_i|$ and they choose to iterate the site that realizes its maximum.

References

[FG2] Fogelman-Soulie, F., E. Goles, G. Weisbuch, *Transient length in Sequential Iteration of Threshold Functions*, Disc. App. Maths., 6, 1983, 95-98.

[G2] Goles, E., *Positive Automata Networks*, in Disordered Systems and Biological Organization, edited by E. Bienenstock, F. Fogelman-Soulie, G. Weisbuch, NATO ASI, Series, F20, Springer Verlag, 1986, 101-112.

[G8] Goles, E., *Lyapunov Functions Associated to Automata Networks*, in Automata Networks in Computer Science, F. Fogelman, Y. Robert, M. Tchuente eds., Manchester University Press, 1987, 58-81.

[G10] Goles, E., *Antisymmetrical Neural Networks*, Disc. App. Math., 13, 1986, 97-100.

[GFP] Goles, E., F. Fogelman-Soulie, D. Pellegrin, *The Energy as a Tool for the Study of Threshold Networks*, Disc. App. Math., 12, 1985, 261-277.

[GM6] Goles, E., S. Martínez, *Exponential Transient Classes of Symmetric Neural Networks for Synchronous and Sequential Updating*, Preprint, Dep. Ing. Mat., Esc. Ing., U. Chile, 1989.

[GO2] Goles, E., J., Olivos, *The Convergence of Symmetric Threshold Automata*, Inf. and Control, 51(2), 1981, 98-104.

[HaL] Haken, A., M. Luby, *Steepest Descent can take Exponential Time for Symmetric Connection Networks*, Complex Systems 2, 1988, 191-196.

[Ho2] Hopfield, J.J., *Neural Networks and Physical Systems with Emergent Collective Computational Abilities*, Proc. Natl. Acad. Sci. USA., 79, 1982, 2554-2558.

[Po] Pomeau, Y., *Invariant in Cellular Automata*, J. Phys. A17, 1984, L415-L418.

[T2] Tchuente, M., *Sequential Iteration of Parallel Iteration*, Theor. Comp. Sci., 48, 1986, 135-144.

4. UNIFORM ONE AND TWO DIMENSIONAL NEURAL NETWORKS

4.1. Introduction

Periodical behaviour in Uniform Neural Networks has been studied since 1959 by several authors [B,FC,GT2,GT3,Ki1,Ko,Sh1,Sh2,T1,T3]. The studies have been chiefly aimed at modelling sensory or cortical neural systems and diffusion processes in one or two dimensions.

In this chapter we present some of these one- and two-dimensional models and we study their dynamical aspects. In some cases we also give experimental results to exhibit the complexity of the dynamical behaviour.

One of the most interesting results we present is Shingai's theorem, which asserts that the period T of any Uniform one-dimensional Neural Network with fixed boundary condition satisfies $T \leq 4$ [Sh1]. This result rests on the boundary condition. In fact, in Chapter 2 we have built examples of Neural Networks with long periods and whose connection matrices are arbitrarily near to symmetric matrices. But contrary to Shingai's theorem we considered the evolution on a torus, so we had a feedback among the "boundary" cells.

For the proof of Shingai's theorem we are led to study several cases. An important class of the networks the theorem refers to can be symmetrized, so the theory here developed allows their description. This avoids the use of extremely long combinatorial arguments.

We also find necessary conditions under which Uniform multidimensional Networks can be symmetrized. When these hypothesis do not hold we are able to furnish examples with non-bounded periods (with the size of the network).

4.2. One-Dimensional Majority Automata

Consider the one-dimensional cellular space $(\mathbb{Z}, V_0)$, with $V_0 = \{-p, ..., 0, ..., p\}$ the symmetric neighbourhood around 0 of length $2p+1$. The Cellular Automata $\mathcal{A} = (\mathbb{Z}, V_0, Q = \{0,1\}, f)$ evolves according to the majority function f:

$$f : Q^n \to Q, \quad f(x_{-p}, ..., x_p) = \begin{cases} 1 & \text{if } \sum_{j=-p}^{p} x_j \geq p+1 \\ 0 & \text{otherwise} \end{cases} \tag{4.1}$$

So f puts in the central cell the most representative values of its neighbours.

The set of configurations C of the Cellular Automata only includes the finite configurations i.e.:

$$C = \{x = (x_i : i \in \mathbb{Z}) : |\text{supp}x| < \infty\} \quad \text{where supp}x = \{i \in \mathbb{Z} : x_i \neq 0\} \quad (4.2)$$

Since $f(0, ..., 0) = 0$ the evolution F_A acts on C and it is given by:

$$F_A : C \to C \text{ where } (F_A x)_i = f(x_{i-p}, ..., x_{i+p}).$$

The tuple $(\mathbb{Z}, V_0, Q = \{0, 1\}, f)$ is called the Majority one-dimensional Cellular Automata.

Let us define some notation. To any finite configuration $x \in C$ associate the numbers:

$$\min x = \min\{i \in \mathbb{Z} : x(i) = 1\}, \quad \max x = \max\{i \in \mathbb{Z} : x(i) = 1\} \quad (4.3)$$

(if $x = 0^{(\infty)}$ i.e. $x_i = 0$ for any $i \in \mathbb{Z}$ we put $\min x = \infty$, $\max x = -\infty$).
For instance if $x = (x_i : i \in \mathbb{Z})$ is:

$i \in \mathbb{Z}$	< 1	1	2	3	4	5	6	7	8	9	10	> 10
x_i	$0^{(\infty)}$	1	0	1	0	0	1	0	0	1	1	$0^{(\infty)}$

we have $\text{supp}x = \{1, 3, 6, 9, 10\}$, $\min x = 1$, $\max x = 10$. The symbol $0^{(\infty)}$ means all the coordinates are equal to 0 to the left of 1 and to the right of 10.

With the above notation any finite configuration $x \in C$ can be written: $x = 0^{(\infty)} x_{\min x} ... x_{\max x} 0^{(\infty)}$. A relevant property which depends on the particular form of a local rule f is the non-expansiveness condition:

$$\min x \leq \min F_A^t(x) \leq \max F_A^t(x) \leq \max x \quad \text{for any } x \in C, \ t \geq 1 \quad (4.4)$$

Let us prove $\min x \leq \min F_C(x)$. The max case is shown in an analogous way and for $t > 1$ it suffices to iterate this inequality. Note $i = \min x$, pick $j < i$. The vector $(x_{j-p}, ..., x_{j+p})$ contains at least $p + 1$ '0' so $f(x_{j-p}, ..., x_{j+p}) = 0$. Then $(F_A(x))_j = 0$, so $\min F_C(x) \geq i$.

Hence the evolution of any finite configuration $x \in C$ can be studied in the finite set of cells $I_x = \{\min x, ..., \max x\}$ because for any other cell $i \notin I_x$ we have $(F_A^t(x))_i = 0$. This is essential to establish the following result:

Theorem 4.1. [T1,T3] Consider the one-dimensional Majority Cellular Automata $A = (\mathbb{Z}, V_0, Q = \{0, 1\}, f)$. Pick a finite configuration $x \in C$, then the evolution $(F_A^t(x) : t \geq 0)$ converges in a finite number of steps to a fixed point.

Furthermore a configuration $x \in C$ is a fixed point under F_A (i.e. $F_A(x) = x$) iff it is of the form:

$$x = 0^{(\infty)} 1^{(p_1)} 0^{(p_2)} 1^{(p_3)} ... 0^{(p_{s-1})} 1^{(p_s)} 0^{(\infty)} \quad \text{with } p_r \geq p+1 \text{ for } r = 1, ..., s \quad (4.5)$$

where $0^{(q)}$ (respectively $1^{(q)}$) means a block of q consecutive 0's (respectively q consecutive 1's). The case $s = 0$ corresponds to the fixed point $x = 0^{(\infty)}$.

Proof. First let us show that $F_A^t x$ converges in a finite number of steps to either fixed points or cycles of length two. If $x = 0^{(\infty)}$ it is evident so assume $x \neq 0^{(\infty)}$.

Take $I_x = \{\min x \leq i \leq \max x\}$. Define the connection matrix $A = (a_{ij} : i, j \in I_x)$ and $b = (b_i : i \in I_x)$ by:

$$a_{ij} = \begin{cases} 1 & \text{if } |i - j| \leq p \quad \text{and } b_i = p + \frac{1}{2} \\ 0 & \text{otherwise} \end{cases} \quad (4.6)$$

With this notation:

$$(F_A^t(x))_i = \mathbb{1}\left(\sum_{j \in I} a_{ij} x_j - b_i\right), \quad i \in I_x \quad (4.7)$$

The matrix A is symmetric, so Theorem 2.1 implies that the orbits of (4.7) are of period 1 or 2 with respect to F_A. Obviously the convergence occurs in a finite number of steps because the Neural Network (I_x, A, b) is finite.

Before showing that points of period 2 cannot exist we prove that fixed points are those configurations which satisfy (4.5).

Let us show that condition (4.5) is sufficient. If $i \notin I_x$ then $(F_A x)_i = x_i = 0$. Pick $i \in I_x$, so it belongs to a block of p_r symbols equal to x_i. Let $V_i = \{i - p, ..., i + p\}$. We have $|\{j \in V_i : x_j = x_i\}| \geq p + 1$ so $f(x_{i-p}, ..., x_{i+p}) = x_i$. We conclude $(F_A x)_i = x_i$, this means x is a fixed point.

To prove condition (4.5) is also necessary assume x is a fixed point. Let $i_0 \in I_x$ be the first coordinate belonging to a block $i_0, ..., i_0 + r - 1$ of length $r \leq p$ with the same state as that of i_0, this means $x_{i_0-1} \neq x_{i_0} = \cdots = x_{i_0+r-1} \neq x_{i_0+r}$. As $i_0 - 1$ belongs to a block of length $\geq p+1$ possesing the opposite state of x_{i_0} the set $\{j \in V_{i_0} : x_j \neq x_i\}$ contains $\{i_0 - p, ..., i_0 - 1, i_0 + r\}$. Then $f(x_{i_0-p}, ..., x_{i_0+p}) = 1 - x_{i_0}$, so x can not be a fixed point.

Now we shall prove that there do not exist configurations of period $T = 2$. Suppose the contrary, take $x \in C$ such that $F_A x \neq x$, $F_A^2 x = x$. Note:

$$x = 0^{\infty} 1^{p_1} 0^{p_2} ... 0^{p_{s-1}} 1^{p_s} 0^{\infty}$$

If $p_r \geq p + 1$ for any $r = 1, ..., s$ the point x would satisfy $F_A x = x$ which is not the case. Then pick as before i_0 the first coordinate such that $x_{i_0-1} \neq x_{i_0} = \cdots =$

$x_{i_0+r-1} \neq x_{i_0+r}$ with $r \leq p$. As any $i < i_0$ belongs to a block of length $\geq p+1$ we have $(F_A x)_i = x_i$. Now $|\{j \in V_{i_0} : x_j \neq x_i\}| \geq p+1$ so $(F_A x)_{i_0} = 1 - x_{i_0}$. This means $(F_A x)_{i_0} = (F_A x)_{i_0-1} = \cdots = (F_A x)_{i_0-(p+1)} = 1 - x_{i_0}$ (because the block of $i_0 - 1$ is of length $\geq p + 1$), so:

$$(F_A^2 x)_{i_0} = (F_A x)_{i_0} = 1 - x_{i_0} \neq x_{i_0}$$

which contradicts the hypothesis $F_A^2 x = x$. We have shown the theorem. ■

The study conducted in Chapter 3 on transient lengths of finite Neural Networks allows finding bounds for the one-dimensional Majority Cellular Automata.

Take $x \in C$ and denote by $\bar{x}$ its restriction to I_x, i.e. $\bar{x}_i = x_i$ for $i \in I_x$. From (4.7) the transient length $\tau(x)$ of x under transformation F_A is the same as the one of $\bar{x}$ evolving in the Neural Network $\mathcal{N}_x = (I_x, A, b)$ with A, b given by (4.6), A being an integer symmetric matrix, the bound (3.13) obtained in Proposition 3.5 holds:

$$\tau(x) \leq \frac{1}{2}(\|A\|_1 + 3\|2b - A\bar{1}\|_1 - n_x) \text{ where } n_x = \max x - \min x + 1 \qquad (4.8)$$

Put $\min x = 1$. Then row i of the matrix A defined in (4.6) is of the following form:
  - if $i \leq p$ then $a_{ij} = 1$ for $j = 1, ..., p + i$; $a_{ij} = 0$ otherwise;
  - if $p < i \leq n_x - p$ then $a_{ij} = 1$ for $j = i - p, ..., i + p$; $a_{ij} = 0$ otherwise;
  - if $i > n_x - p$ then $a_{ij} = 1$ for $j = p + 1 + (n_x - i), ..., n_x$; $a_{ij} = 0$ otherwise.

Hence $\|A\|_1 = 2 \sum_{r=1}^{p} (p + r) + (2p + 1)(n_x - 2p) = (2p + 1)n_x - p(p + 1)$ and

$$\|2b - A\bar{1}\|_1 = 2 \sum_{r=1}^{p} |2p + 1 - (p + r)| + (n_x - 2p)|2p + 1 - (2p + 1)| = p(p + 1).$$

Hence $\tau(x) \leq \frac{1}{2}((2p + 1)n_x - p(p + 1)n_x)$. We have obtained:

Proposition 4.1. In the one-dimensional Majority Cellular Automata $\mathcal{A} = (\mathbb{Z}, V_0 = \{-p, \cdots, p\}, \{0, 1\}, f)$ the transient length $\tau(x)$ of any finite configuration $x \in C$ is bounded by:

$$\tau(x) \leq p(p + n_x + 1) \quad ■ \qquad (4.9)$$

A sharper bound can be established:

Proposition 4.2. [T1] Consider the one-dimensional Majority Cellular Automata. Let $x \in C$ be a finite configuration. Decompose its support in the blocks of consecutive 1's, we denote by $r(x)$ the cardinality of this set of blocks:

$$\text{supp} x = \bigcup_{k=1}^{r(x)} \{j_k \leq i \leq \ell_k\} \text{ with } j_k \leq \ell_k < j_{k+1} \leq \ell_{k+1} \text{ for } 1 \leq k \leq r(x) \qquad (4.10)$$

Then:

$$r(F_A(x)) \le r(x) \quad \text{and} \quad r(F_A(x)) = r(x) \quad \text{iff } x \text{ is a fixed point} \qquad (4.11)$$

Hence the transient length $\tau(x)$ satisfies $\tau(x) \le r(x)$. Moreover, if $\tilde{x}$ is the fixed point at which $F_A^t(x)$ converges we have:

$$\tau(x) \le r(x) - r(\tilde{x}) \qquad (4.12)$$

Proof. Pick $x \in \mathcal{C}$, denote $r = r(x)$. With the above notation we have
$x_{j_k} = \cdots = x_{\ell_k} = 1$ and $x_j = 0$ for $j \notin \bigcup_{k=1}^{r} \{j_k \le i \le \ell_k\}$. Also decompose
$\text{supp} F_A(x) = \bigcup_{k=1}^{r'} \{j_k' \le i \le \ell_k'\}$ with $j_k' \le \ell_k' < j_{k+1}' \le \ell_{k+1}'$.

By definition $(F_A(x))_{j_k'-1} = 0$, $(F_A(x))_{j_k'} = 1$, $F_A(x))_{\ell_k'} = 1$,
$(F_A(x))_{\ell_k'+1} = 0$, which is equivalent to the inequalities:

$$\sum_{i=-p}^{p} x_{j_k'-1+i} \le p, \quad \sum_{i=-p}^{p} x_{j_k'+i} \ge p+1, \quad \sum_{i=-p}^{p} x_{\ell_k'+i} \ge p+1, \quad \sum_{i=-p}^{p} x_{\ell_k'+1+i} \le p.$$

Hence $x_{j_k'+p} - x_{j_k'-1-p} > 0$, $x_{\ell_k'-p} - x_{\ell_k'+p+1} < 0$, which necessarily implies:
$x_{j_k'+p} = x_{\ell_k'-p} = 1$ and $x_{j_k'-p-1} = x_{\ell_k'+p+1} = 0$.

For any $k \in \{1, ..., r'\}$ we have $x_{j_k'+p} \in \text{supp} x$ so there exists a unique integer
$\phi(k) \in \{1, ..., r\}$ such that $j_{\phi(k)} \le j_k' + p \le \ell_{\phi(k)}$. If $1 \le k < m \le r'$ we have
$j_k' + p < \ell_k' + p + 1 \le j_m' + p$. As $x_{j_k'+p} = x_{j_m'+p} = 1$ and $x_{\ell_k'+p+1} = 0$ the function
ϕ is one-to-one and increasing. So $r' \le r$.

Now suppose $r' = r$. As ϕ is an onto one-to-one increasing function we get
$\phi(k) = k$ for any $k \in \{1, ..., r\}$. Then $j_k \le j_k' + p \le \ell_k$. But $\ell_k' - p$ also belongs to
$\text{supp} x$, denote $\psi(k)$ the unique number in $\{1, ..., r\}$ such that $j_{\psi(k)} \le \ell_k' - p \le \ell_{\psi(x)}$.
By the same analysis as before ψ is increasing one-to-one so $\psi(k) = k$. This implies
$j_k \le \ell_k' - p \le \ell_k$.

Then $\ell_k - j_k \ge |\ell_k' - p - j_k' - p| = |(\ell_k' - j_k') - 2p|$. If $\ell_k' - j_k' < p$ we deduce
$\ell_k - j_k \ge p+1$. By the analysis in Theorem 4.1, the coordinates $j_k \le i \le \ell_k$ are
fixed i.e. $(F_A x)_i = x_i = 1$ for $j_k \le i \le \ell_k$. This necessarily implies $\ell_k' - j_k' \ge p+1$.
But $r' = r$ implies that $j_k' = j_k$, $\ell_k' = \ell_k$. From Theorem 4.1 we conclude x is a
fixed point. So (4.11) is satisfied. This implies that the transient length $\tau(x)$ of
any point $x \in \mathcal{C}$ satisfies (4.12). ∎

Concerning bound (4.12), there exist examples where the inequality is strict
while for others it becomes an equality. For instance take:

$$x = 0^{(\infty)} 11101110110001110^{(\infty)}$$

Its evolution is:

$$F_A x = 0^{(\infty)} 01111111100110000^{(\infty)}$$

$$F_A^2 x = 0^{(\infty)} 01111111111000000^{(\infty)}$$

The latter configuration $\tilde{x} = F_A^2 x$ is a fixed point. We have $r(x) = 4$, $r(F_A x) = 2$, $r(\tilde{x}) = 1$. The transient lengths are $\tau(x) = 2$, $\tau(F_A x) = 1$. So $\tau(x) < r(x) - r(\tilde{x}) = 3$, $\tau(F_A x) = r(F_A x) - r(\tilde{x}) = 1$.

4.3. Two-dimensional Majority Cellular Automata

On $\mathbb{Z}^2$ take the von-Neumann neighbourhood:

$$V_0 = V_0^N = \{(-1,0),(1,0),(0,0),(0,1),(0,-1)\}$$

We shall study the Cellular Automata $A = (\mathbb{Z}^2, V_0, Q = \{0,1\}, f)$ where the local function $f : Q^{|V_0|} \to Q$ depends on a parameter $\theta \in \{1, ..., |V_0| - 1\}$ and has the form:

$$f(x_j : j \in V_0) = \mathbb{1}\Big(\sum_{j \in V_0} x_j - \theta \Big) \tag{4.13}$$

We call A a θ-threshold two-dimensional Cellular Automata. Its evolution is given by:

$$(F_A(x))_i = f(x_j : j \in V_i)$$

Denote by $0^{(\infty)}$ (respectively $1^{(\infty)}$) the constant configuration equal to 0 (respectively 1). As $1 \leq \theta \leq 4$ the points $0^{(\infty)}$ and $1^{(\infty)}$ are fixed. For $\theta = 1$ the dynamical behaviour is extremely simple because any configuration $x \neq 0^{(\infty)}$ converges to $1^{(\infty)}$ and for $\theta = 4$ the opposite occurs, every $x \neq 1^{(\infty)}$ converges to $0^{(\infty)}$. We shall study the other cases $\theta = 3$ or $\theta = 2$.

Recall that our configuration space is $C = \{x \in \{0,1\}^{\mathbb{Z}^2} : |\operatorname{supp} x| < \infty\}$. As $f(0,0,0,0,0) = 0$ the transformation F_A acts on C.

For its analysis let us introduce some notation. For any bounded set $C \subset \mathbb{Z}^2$ its 1-projection is $C^{(1)} = \{i_1 \in \mathbb{Z} : \exists i_2 \in \mathbb{Z} \text{ such that } (i_1, i_2) \in C\}$, and its 2-projection is $C^{(2)} = \{i_2 \in \mathbb{Z} : \exists i_1 \in \mathbb{Z} \text{ such that } (i_1, i_2) \in C\}$.

Now take a finite configuration $x \in C$. We denote $\min^{(\ell)} x = \min(\operatorname{supp} x)^{(\ell)}$, $\max^{(\ell)} x = \max(\operatorname{supp} x)^{(\ell)}$ for $\ell = 1, 2$. Then the intervals

$$I_x^{(1)} = \{\min^{(1)} x \leq i_1 \leq \max^{(2)} x\}, \ I_x^{(2)} = \{\min^{(2)} x \leq i_2 \leq \max^{(2)} x\}$$

contain the leftmost and the rightmost sites of $\operatorname{supp} x$ with respect to the first and the second coordinate respectively. By construction the rectangle $R_x = I_x^{(1)} \times I_x^{(2)}$ is the smallest rectangle in $\mathbb{Z}^2$ containing $\operatorname{supp} x$ (see Figure 4.1)

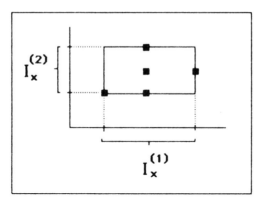

Figure 4.1. Rectangle generated by a configuration.

We observe that the supports of the iterated points remain in the initial rectangle; this means F_A verifies the non-expansiveness condition:

$$\text{supp}(F_A^t(x)) \subset R_x \quad \text{for any } t \geq 0 \qquad (4.14)$$

Then we can study the evolution of x in the sites of $R(x)$, since in any other site $i \notin R_x$ we have $(F_A^t x)_i = 0$.

As we did in the one dimensional case, we can study the evolution of $F_A^t x$ by using the results of finite Neural Networks. Take $I = R_x$, $A = (a_{ij} : i, j \in I)$ the symmetric matrix and $b = (b_i : i \in I)$ the threshold vector defined by:

$$a_{ij} = \begin{cases} 1 & \text{if } j \in V_i \\ 0 & \text{if not} \end{cases} \quad \text{and} \quad b_i = \theta - \tfrac{1}{2} \qquad (4.15)$$

Denote $\bar{x}$ the restriction of x to the sites I and $F_N(\bar{x}) = \mathbb{1}(A\bar{x} - b)$ the evolution of the Neural Network $N = (I, A, b)$. Then $(F_A(x))_i = (F_N(\bar{x}))_i$ for any $i \in I$. As A is symmetric the result of Theorem 2.1 holds, so $F_A^t x$ converges to a fixed point or a period-2 orbit. In what follows we shall not distinguish between $x \in C$ and $\bar{x} \in \{0,1\}^{R_x}$.

4.3.1. 3-Threshold Case. Assume $\theta = 3$. Then $f(x_j : j \in V_0) = \mathbb{1}(\sum_{j \in V_0} x_j - 3)$ is the majority rule, it assigns to a tuple the most representative value. So A is the two-dimensional Majority Cellular Automata.

Unlike in the one dimensional case, for the majority local rule on $\mathbb{Z}^2$ we have fixed points as well as period-2 orbits. For instance: $x = \begin{matrix} 0 & 0 \\ 0 & 0 \end{matrix}$ is a fixed point and:

$$x = \begin{matrix} 1 & 1 & 1 & 0 & 1 & 0 & 1 & 1 \\ 1 & 1 & 0 & 1 & 0 & 1 & 1 & 1 \end{matrix}$$

is a period 2 configuration. In fact its iterate:

$$F_A(x) = \begin{matrix} 1 & 1 & 0 & 1 & 0 & 1 & 1 & 1 \\ 1 & 1 & 1 & 0 & 1 & 0 & 1 & 1 \end{matrix}$$

is different from x and $F_A^2(x) = x$.

More complex fixed points for the majority rule are:

$$\begin{matrix} 1 & 1 & 1 & 1 & 1 \\ 1 & 0 & 0 & 0 & 1 \\ 1 & 0 & 0 & 0 & 1 \\ 1 & 1 & 1 & 1 & 1 \end{matrix} \qquad \begin{matrix} 1 & 1 \\ 1 & 1 & 1 & 1 & 1 & 1 & 1 \\ & & & & 1 \\ 1 & 1 & 1 & 1 & 1 & 1 \\ & & & & 1 & 1 \end{matrix}$$

$$\begin{matrix} 1 & 1 & & & & & & 1 & 1 \\ 1 & 1 & 1 & 1 & 1 & 1 & 1 & 1 & 1 \\ 0 & 0 & 0 & 0 & 0 & 0 & 0 & 0 & 0 \\ 1 & 1 & 1 & 1 & 1 & 1 & 1 & 1 & 1 \\ 1 & 1 & & & & & & 1 & 1 \end{matrix}$$

Forbidden structures for fixed points are constituted by small holes, for instance:

$$\begin{matrix} 1 & 1 \\ 1 & 0 \\ 1 & 1 \end{matrix} \qquad \text{or} \qquad \begin{matrix} 0 & 0 \\ 0 & 1 \\ 0 & 0 \end{matrix}$$

As we did in the one-dimensional case, we can bound the transient length $\tau(x)$ by expression (3.13) of Proposition 3.5. Then:

$$\tau(x) \le \frac{1}{2}(\|A\|_1 + 3\|2b - A\bar{1}\|_1 - |R_x|)$$

For $R_x = I_x^{(1)} \times I_x^{(2)}$ write $n_x^{(\ell)} = $ length of $I_x^{(\ell)} = \max^{(\ell)} x - \min^{(\ell)} x + 1$, then $|R_x| = n_x^{(1)} n_x^{(2)}$. From the form of A, b in (4.15) we find:

$$\|A\|_1 = 5|R_x| - 2n_x^{(1)} - 2n_x^{(2)}, \quad \|2b - A\bar{1}\| = n_x^{(1)} + n_x^{(2)} + 4,$$

so $\frac{1}{2}(\|A\|_1 + 3\|2b - A\bar{1}\|_1 - |R_x|) = 2|R_x| + \frac{n_x^{(1)} + n_x^{(2)}}{2} + 6$. The Perimeter of R_x being $2(n_x^{(1)} + n_x^{(2)}) - 4$ we have obtained:

Proposition 4.3. In the two-dimensional Majority Cellular Automata the orbits are of length ≤ 2. The transient length $\tau(x)$ of any finite configuration $x \in C$ is bounded by:

$$\tau(x) \le \Gamma_3(R_x) = 2|R_x| + \frac{1}{4} \text{ Perimeter } R_x + 7 \qquad (4.16)$$

where R_x is the smallest rectangle containing supp$x$. ■

4.3.2. 2-Threshold Case. Assume $\theta = 2$ for the local function (4.13). Thus any cell needs only two cells in state 1 to be activated. We have $\|2b - A\bar{1}\| = 2|R_x| - 3(n_x^{(1)} + n_x^{(2)}) + 8$. So $\frac{1}{2}(\|A\|_1 + 3\|2b - A\bar{1}\|_1 - |R_x|) = 5|R_x| - \frac{11}{2}(n_1^{(x)} + n_2^{(x)}) + 36$.

Proposition 4.4. In the 2-threshold two-dimensional Cellular Automata the orbits are of length ≤ 2. The transient length of any finite configuration $x \in C$ is bounded by:

$$\tau(x) \leq \Gamma_2(R_x) = 5|R_x| - \frac{11}{4} \text{ Perimeter } R_x + 37 \quad ■ \qquad (4.17)$$

Obviously the last bound can be sharpened if we can partition R_x into sub-rectangles separated by a distance > 1. More precisely assume $(R_x^{(\ell)} : \ell = 1, ..., r)$ are disjoint rectangles included in R_x and satisfying supp$x \subset \bigcup_{\ell=1}^{r} R_x^{(\ell)} \subset R_x$ and

$$d_1(R_x^{(\ell)}, R_x^{(\ell')}) = \inf\{d_1(i, i') : i \in R_x^{(\ell)}, i' \in R_x^{(\ell')}\} > 1 \quad \text{for } \ell \neq \ell';$$

where $d_1(i, i') = |i_1 - i'_1| + |i_2 - i'_2|$. Then the transient length of x is bounded by: $\tau(x) \leq \max\{\Gamma_2(R_x^{(\ell)}) : \ell = 1, ..., r\}$.

Now we shall describe the patterns in steady state.

Proposition 4.5. Consider the 2-threshold two-dimensional Cellular Automata.

(i) A configuration $x \in C$ is a fixed point iff supp$x$ is a union of disjoint rectangles at a distance > 2 from one another: supp$x = \bigcup_{\ell \in L} R_\ell$ with $d_1(R_\ell, R_{\ell'}) > 2$ for every $\ell \neq \ell'$.

(ii) Assume that $x \in C$ is a point of period 2 in which each cell of $C_x = \text{supp}x \cup \text{supp}F_A x$ is of period 2. Then supp$x \cap \text{supp}F_A x = \phi$ and C_x is a union of disjoint rectangles, at a distance > 1 from one another: $C_x = \bigcup_{\ell \in L} R_\ell$ with $d_1(R_\ell, R_{\ell'}) > 1$ for every $\ell \neq \ell'$. Furthermore if $i \in R_\ell$, $j \in R_{\ell'}$ with $\ell \neq \ell'$, are such that $d_1(i, j) = 2$ then $x_i = 1$ iff $x_j = 0$.

Proof. (i) For any two points $i = (i_1, i_2)$, $j = (j_1, j_2)$ consider the intervals $I_\ell = \{i'_\ell : \min(i_\ell, j_\ell) \leq i'_\ell \leq \max(i_\ell, j_\ell)\}$ for $i = 1, 2$. By $R(i, j) = I_1 \times I_2$ we denote the rectangle generated by (i, j).

Now note that if $i, j \in \text{supp}x$ and $d_1(i, j) \le 2$ then the rectangle $R(i, j)$ determined by these points is contained in $\text{supp}x$. This occurs because $\theta = 2$. The two non trivial cases occur when $d_1(i, j) = 2$; as shown in Figure 4.2.

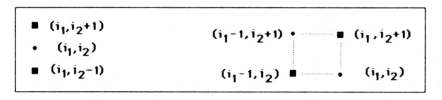

Figure 4.2. Rectangles generated by points for $\theta = 2$.

Now suppose that the line $L = \{i_1 \le i'_1 \le j_1\} \times \{i_2\}$ is contained in $\text{supp}x$ for $i_1 \le j_1$, i_2 fixed. Then if $k = (k_1, k_2) \in \text{supp}x$ is such that $d_1(k, L) \le 2$ we deduce that the rectangle $R(k, L)$ generated by k, L is also contained in $\text{supp}x$. The non-trivial case occurs when $d_1(k, L) = 2$, is illustrated in Figure 4.3.

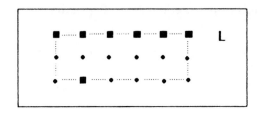

Figure 4.3. Rectangle generated by lines and points for $\theta = 2$.

With the two above remarks we easily deduce the assertion of part (i).

(ii) Denote $C = \text{supp}x \cup \text{supp}F_A x$. For any pair $i, i' \in C$ we write $i \approx i'$ iff $i' \in V_i$. By $C(i)$ we mean the equivalent class of i for the equivalence relation $\approx$. From definition $j \in C$ and $j \notin C(i)$ implies $d_1(i, j) > 1$, so $d_1(C - C(i), C(i)) > 1$. Then it suffices to show that $C(i)$ is a rectangle, so we may assume $C(i) = C$.

Recall that the hypothesis means that at any site $i \in C$ the configuration x is of period 2 : $(F_A^2 x)_i = x_i \ne (F_A x)_i$. Now take $i \in \text{supp}x$, then any $j \in V_i - \{i\}$ satisfies $x_j = 0$. In fact the contrary would imply $(F_A x)_i = 1 = x_i$ which is a contradiction. Also if $i \in \text{supp}F_A x$ we have $(F_A x)_j = 0$ for $j \in V_i - \{i\}$.

On the other hand, $i \in C$ implies that at least two points of $V_i - \{i\}$ must belong to C. In fact as $x_i = 0$ or $(F_A x)_i = 0$ the contrary would imply that $(F_A^t x)_i = 0$ for $t \ge 2$ contradicting $i \in C$.

Furthermore, if the elements (i_1, i_2), $(i_1 - 1, i_2)$, $(i_1, i_2 + 1)$ belong to C then $(i_1 - 1, i_2 + 1) \in C$. In fact by the above analysis we may assume $x_{(i_1 - 1, i_2)} =$

$x_{(i_1,i_2+1)} = 1$ (if not so we pick $F_A x$ instead of x) which implies
$(F_A x)_{(i_1-1,i_2+1)} = 1$ so $(i_1 - 1, i_2 + 1) \in C$. This is called the corner argument.
Analogous implications are obtained for other triples, for instance
$(i_1, i_2), (i_1 + 1, i_2), (i_1, i_2 + 1) \in C$ implies $(i_1 + 1, i_2 + 1) \in C$.

Let $k^{(1)} = \inf C^{(1)}$ be the leftmost element of the first coordinate of C.
Pick $m^{(2)} = \max\{i_2 \in \mathbb{Z} : (k^{(1)}, i_2) \in C\}$. By definition $(k^{(1)} - 1, m^{(2)})$
and $(k^{(1)}, m^{(2)} + 1)$ do not belong to C; then necessarily $(k^{(1)}, m^{(2)} - 1)$ and
$(k^{(1)} + 1, m^{(2)})$ belong to C. Now define

$$m^{(1)} = \max\{j_1 \in \mathbb{Z} : (i_1, m^{(2)}) \in C \quad \text{for any } k^{(1)} \le i_1 \le j_1\} \quad \text{and}$$

$$k^{(2)} = \min\{j_2 \in \mathbb{Z} : (m^{(1)}, i_2) \in C \quad \text{for any } j_2 \le i_2 \le m^{(2)}\}.$$

We must prove $C = I^{(1)} \times I^{(2)}$ where $I^{(\ell)} = \{k^{(\ell)} \le i_\ell \le m^{(\ell)}\}$ for $\ell = 1, 2$ (see
Figure 4.4).

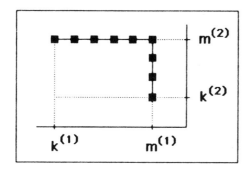

Figure 4.4. Rectangle exhibiting a two-period behaviour.

Since $(k^{(1)}, m^{(2)} - 1), (i_1, m^{(2)})$ belong to C for any $k^{(1)} \le i_1 \le m^{(1)}$ we can
use the corner argument to deduce that $(i_1, m^{(2)} - 1)$ belong to C for everyone of
these i_1. If $k^{(2)} < m^{(2)} - 1$ from definition we have $(m^{(1)}, m^{(2)} - 2) \in C$. Then the
corner argument applied to this point and to $(i_1, m^{(2)} - 1)$ for $k^{(1)} \le i_1 \le m^{(1)}$
implies $(i_1, k^{(2)} - 2)$ belongs to C for all such i_1. Thus we proved $C \subset I^{(1)} \times I^{(2)}$.

As C is closed for the equivalence relation $\approx$ the proof is complete when we
show $i = (i_1, i_2) \notin C$ for $k^{(1)} \le i_1 \le m^{(1)}$ and $i_2 \in \{m^{(2)} + 1, k^{(2)} - 1\}$ or
$i_1 \in \{k^{(1)} - 1, m^{(2)} + 1\}$ and $k^{(2)} \le i_2 \le m^{(2)}$. The case $i_1 = k^{(1)} - 1$ follows from
the definition of $k^{(1)}$.

The other three cases admit a proofs which are analogous among them. For
instance consider $k^{(1)} \le i_1 \le m^{(1)}$ and $i_2 = k^{(2)} - 1$ suppose

$$D = \{k^{(1)} \le i_1 \le m^{(1)} : (i_1, k^{(2)} - 1) \in C\} \ne \phi.$$

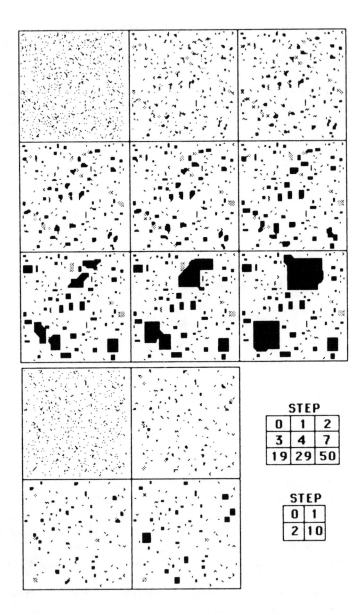

Figure 4.5. Dynamics of the 2-threshold two-dimensional Cellular Automata in a 95×95 torus. (i) Pattern evolving to the fixed point '1'. (ii) A typical two-periodic steady state with cycles and fixed rectangles of 1's.

Since $(i_1, k^{(2)}) \in C$ for any $k^{(1)} \leq i_1 \leq m^{(1)}$ the corner argument implies $(m^{(1)}, k^{(2)} - 1) \in C$, which contradicts the definition of $k^{(2)}$. Then the result follows. ∎

Hence the stable configurations of 2-threshold two-dimensional Cellular Automata consist of conviniently separated combinations of patterns of form (i), (ii) of Proposition 4.5.

Typical dynamics and fixed points of this network are given in Figures 4.5 and 1.9.

4.4. Non-Symmetric One-Dimensional Bounded Neural Networks

As usual the finite set of cells will be $I = \{1, ..., n\}$ with its canonical order. Note $V_i = \{i - 1, i, i + 1\}$. The neighbour of each site i is $V_i' = V_i \cap I$ so the extremal cells posses only two neighbours.

Then the graph $G = (I, V')$ is of the form:

$$(i, j) \in V' \quad \text{iff } j \in V_i'$$

The matrix of connection weights $A = (a_{ij} : i, j \in I)$ will be uniform but not necessarily symmetric, so:

$$a_{ij} = 0 \text{ if } (i, j) \notin V', \ a_{i,i-1} = a, \ a_{i,i} = c, \ a_{i,i+1} = d \qquad (4.18)$$

The threshold θ will be the same for every site. In this context previous automaton is called a one-dimensional Bounded Neural Networks (BNN).

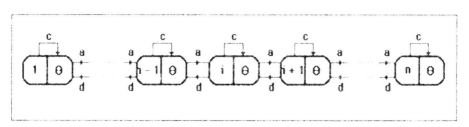

Figure 4.6. One-dimensional Bounded Neural Network.

This Bounded Neural Network is denoted by $\mathcal{N} = (I, a, c, d, \theta)$. Its evolution is given by:

$$F_{\mathcal{N}}(x) = \mathbb{1}(Ax - \bar{\theta}) \quad \text{for } x \in \{0, 1\}^n \qquad (4.19)$$

where $\bar{\theta}$ is the constant vector equal to θ and A is the $n \times n$ matrix given by (4.18).

Also for our analysis it will be useful to write the above evolution in a local form. Add 0 and $n + 1$ to the set of cells and assume that for any trajectory $\{x(t)\}_{t \geq 0}$ the states of these new cells remain fixed at 0, i.e. $x_0(t) = x_{n+1}(t) = 0$ for any $t \geq 0$. Thus $x(t)$ does only evolve in the sites $i \in I = \{1, ..., n\}$. The evolution (4.19) can be written in terms of a local function $f : \{0, 1\}^3 \to \{0, 1\}$:

$$\text{for } i \in I : (F_{\mathcal{N}} x)_i = f(x_j : j \in V_i) \quad \text{with}$$
$$f(u_1, u_2, u_3) = \mathbb{1}(au_1 + cu_2 + du_3 - \theta) \tag{4.20}$$

The evolution of Bounded Neural Networks was studied by Shingai in [Sh2], a paper which is extremely difficult to read and put in a rigorous language. He proved that their period T is bounded by 4, using combinatorial arguments, to treat the different cases.

In what follow we will show Shingai's theorem in a more simple way. In fact, some particular cases can be symmetrized, which immediately implies $T \leq 2$. This reduces the large amount of work involved in the proof of the theorem.

For the sign function we use the definition $\text{sign}(u) = 1$ if $u > 0$, -1 if $u < 0$ and 0 if $u = 0$. We shall be able to put $\mathcal{N}$ into equivalence with a symmetrical network when $\text{sign}(a) = \text{sign}(d)$, which obviously covers the symmetrical case $a = d$.

Lemma 4.1. Assume $\text{sign}(a) = \text{sign}(d)$. Then the one-dimensional Bounded Neural Network $\mathcal{N} = (I, a, c, d, \theta)$ is equivalent to a symmetric Neural Network $\mathcal{N} = (I, A', b')$ in the following strong sense:

$$\mathbb{1}(Ax - \bar{\theta}) = \mathbb{1}(A'x - b') \quad \text{for any } x \in \{0, 1\}^n$$

Then all the dynamical properties of $\mathcal{N}$ are those of a symmetric Neural Network, in particular the orbits of its synchronous updating are of period $T \leq 2$.

Proof. If $a = d = 0$ the matrix A is symmetric and then there is nothing to be proved. Let $a \neq 0$, $d \neq 0$. Define $a'_{ij} = (\frac{d}{a})^i a_{ij}$ and $b'_i = (\frac{d}{a})^i \theta$. Note that $\bar{A}$ is symmetric, in fact $a_{i,i-1} = \frac{d^i}{a^{i-1}} = a_{i+1,i}$ for each $1 < i < n$. By definition we have $(A'x - b')_i = (\frac{d}{a})^i (Ax - \bar{\theta})_i$. As $(\frac{d}{a})^i > 0$ we deduce $\mathbb{1}(Ax - \bar{\theta}) = \mathbb{1}(A'x - b')$
∎

Before studying the other cases note that the finiteness of the network implies that we can always consider the relation $au_1 + cu_2 + du_3 - \theta \neq 0$ to hold for each $(u_1, u_2, u_3) \in \{0, 1\}^3$. In fact the θ value can always be perturbed conveniently.

On the other hand, if $a = 0$, $d \neq 0$ (or $a \neq 0$, $d = 0$) we can always perturb coefficient a (respectively d) in order that $\text{sign}(a) = \text{sign}(d)$ without changing the dynamics of the Neural Network. So in these latter cases the relation $T \leq 2$ is also satisfied.

Then consider $\text{sign}(a) \neq \text{sign}(d)$, $a \neq 0 \neq d$. Since the boundary condition $x_0(t) = x_{n+1}(t) = 0$ for any $t \geq 1$ is symmetric with respect to the inversion $i \to i' = n + 1 - i$ and in this new coordinates i' we have $a' = d$, $d' = a$, $c' = c$, we can always assume without loss of generality that $d < 0 < a$.

Lemma 4.2. Let $\theta > 0$ and $d < 0 < a$. Then the only periodic orbits of the Bounded Neural Network $\mathcal{N} = (I, a, c, d, \theta)$ are fixed points, i.e. $T = 1$.

Proof. From $\theta > 0$ the point $x = (0, ..., 0)$ is fixed.

Suppose $a - \theta < 0$. This implies $a + d - \theta < 0$ so $f(\cdot, 0, \cdot) = 0$ which means $\{i \in I : x_i(t) = 0\} \subset \{i \in I : x_i(t+1) = 0\}$. Then every initial condition converges to a fixed point.

Analogously if $c - \theta < 0$ we get $c + d - \theta < 0$, so $f(0, \cdot, \cdot) = 0$, whence we infer that every point converges to $(0, ..., 0)$.

Then suppose $a - \theta > 0$, $c - \theta > 0$. If $d + c - \theta > 0$ we have $f(\cdot, \cdot, \cdot) = 1$ for any tuple $(\cdot, \cdot, \cdot) \neq (0, 0, 0)$. Hence the evolution of any initial condition different from $(0, \cdots, 0)$ converges to the fixed point $(1, \cdots, 1)$. So we can also assume $d + c - \theta < 0$.

The above contraints imply $f(0, \cdot, 1) = 0$. Let $x(0)$ be an initial condition different from $(0, \cdots, 0)$, denote $i_0 = \min(\text{supp}\, x(0))$. From $f(0, \cdot, 1) = 0$ we deduce $x_j(t) = 0$ for every $j < i_0$ and any $t \geq 1$.

Assume $x_{i_0+1}(0) = 1$; then $x_{i_0}(1) = 0$ and so $x_{i_0}(t) = 0$ for any $t \geq 1$. On the other hand, if $x_{i_0+1}(0) = 0$ and $x_{i_0+2}(0) = 0$ we get $x_{i_0}(1) = 1$ but $x_{i_0}(t) = 0$ for $t \geq 2$. In the other case, $x_{i_0+1}(0) = 0$, $x_{i_0+2}(1) = 1$, we have $x_{i_0+1}(1) = f(1, 0, 1)$. If $a + d - \theta > 0$, we deduce $x_{i_0+1}(1) = 1$ so $x_{i_0}(t) = 0$ for any $t \geq 2$. Then it is easy to show any initial condition converges to the fixed point $(0, \cdots, 0)$.

If $a + d - \theta < 0$ we have $x_{i_0}(0) = x_{i_0}(1) = x_{i_0}(2) = 1$, $x_{i_0+1}(0) = x_{i_0+1}(1) = 0$, $x_{i_0+2}(0) = 1$. By the above analysis and the equality $f(\cdot, \cdot, 0) = 1$ we deduce the points of the form

$$x(0) = (\quad \underset{\underset{i=1}{\uparrow}}{0}, \quad \cdots \quad 0, \quad \underset{\underset{i=i_0}{\uparrow}}{1}, \quad 0, \quad 1, \quad 0, \quad \cdots \quad 0, \quad \underset{\underset{i=n}{\uparrow}}{1} \quad)$$

are fixed and any other initial condition converges to $(0, \cdots, 0)$ (by parity the allowable values for i_0 are of the form $i_0 = n - 2k$).

We have analyzed all the possible cases, hence the result holds. ∎

Now assume $\theta < 0$. As $d < 0 < a$ we have $a - \theta > 0$. We shall examine other cases according to the sign of the parameters $d - \theta$, $c - \theta$, $d + c - \theta$, $a + d - \theta$, $a + c - \theta$, $a + d + c - \theta$ (this covers all the posibilities of $\mathbb{1}(ax_1 + cx_2 + dx_3 - \theta)$ for $(x_1, x_2, x_3) \in \{0, 1\}^3$). But there exist some restrictions about the signs of these coefficients. In fact $u_1 d + u_2 c - \theta > 0$ implies $a + u_1 d + u_2 c - \theta > 0$ for $(u_1, u_2) = (1, 0), (0, 1)$ or $(1, 1)$. Also $v_1 a + v_2 c - \theta < 0$ implies $v_1 a + d + v_2 c - \theta < 0$ for $(v_1, v_2) = (0, 1)$ or $(1, 1)$. The last restriction results from considering the equality $(a + d + c - \theta) - \theta = (a + d) - \theta + c - \theta = (a + c) - \theta + d - \theta$. As $\theta < 0$ we can deduce from $(a + d + c - \theta) > 0$ that $a + d - \theta > 0$ or $a + c - \theta > 0$.

By taking into account the above implications, the number of cases to be studied reduces to 17 instead of the original $64 = 2^6$ ones. The table of signs for these cases is as follows:

Case	1	2	3	4	5	6	7	8	9
Parameter									
$d - \theta$	+	+	+	−	−	−	−	−	−
$c - \theta$	+	−	−	+	+	+	+	−	−
$d + c - \theta$	+	−	−	+	+	−	−	−	−
$a + d - \theta$	+	+	+	+	−	−	−	+	−
$a + c - \theta$	+	+	−	+	+	+	+	−	−
$a + d + c - \theta$	+	+	−	+	+	+	−	−	−
Period T	$= 1$	≤ 2	$= 2$	$= 1$	$= 1$	$= 1$	$= 1$	≤ 2	≤ 2

Case	10	11	12	13	14	15	16	17
Parameter								
$d - \theta$	+	+	−	+	−	−	−	−
$c - \theta$	+	−	+	+	+	−	−	−
$d + c - \theta$	−	−	−	−	−	−	−	−
$a + d - \theta$	+	+	+	+	+	−	+	+
$a + c - \theta$	+	+	+	+	+	+	+	+
$a + d + c - \theta$	−	−	+	+	−	−	+	−
Period T	≤ 2	≤ 2	$= 4$	$= 3$	$= 3$	$= 3$	$= 3$	$= 3$

Table 4.1. Signs and periods in the case $\theta < 0$, $d < 0 < a$. '+' codes > 0 and '−' codes < 0.

For each case $\ell \in \{1, \cdots, 17\}$ on the above table we denote by f_ℓ its local function $f_\ell : \{0, 1\}^3 \to \{0, 1\}$.

Some of these cases can be analyzed by putting them into equivalence with symmetrical Neural Networks.

Lemma 4.3. Let $\theta < 0$, $d < 0 < a$. Then the Bounded Neural Network $\mathcal{N} = (I, a, c, d, \theta)$ associated to f_1, f_2, f_3 is equivalent to a symmetrical network, hence its orbits have a period $T \leq 2$.

Proof. It can be easily shown that when we develop the sign contraints for these functions there exists $d > 0$ which satisfies them. Then we can assume $d > 0$, $a > 0$, so Lemma 4.1 implies the result.

Let us show for f_3 that there exists some $d > 0$ satisfying the sign contraints (for f_1, f_2 it is easier). We must have $d < -(c - \theta)$, $d < -(a + c - \theta)$. Since $-(c - \theta)$ and $-(a + c - \theta)$ are strictly positive we can pick $d > 0$. ∎

For f_3 we have $f_3(\cdot, 0, \cdot) = 1$, $f_3(\cdot, 1, \cdot) = 0$ then the network has only orbits of period $T = 2$. For f_1 we can also pin down the result:

Lemma 4.4. Let $\theta < 0$, $d < 0 < a$. The orbits of the Bounded Neural Network $\mathcal{N} = (I, a, c, d, \theta)$ associated to f_1, f_4, f_5, f_6, f_7 are only fixed points, $T = 1$.

Proof. For $\ell = 1, 4, 5$ we have $f_\ell(\cdot, 1, \cdot) = 1$. Then $\operatorname{supp} x(t) \subset \operatorname{supp} x(t + 1)$, so starting from any configuration we arrive at a fixed point.

For $\ell = 6$ we shall use the boundary conditions. Assume that $x(0)$ belongs to an orbit. Assume that coordinate k is fixed for some $1 \leq k \leq n + 1$. We shall prove that coordinate $k - 1$ is also fixed. Suppose $x_k(t) = 0$ for any $t \geq 0$, from $f_6(\cdot, \cdot, 0) = 1$ we deduce $x_{k-1}(t) = 1$ for any $t \geq 0$. If $x_k(t) = 1$ for $t \geq 0$ and for some t_0 we have $x_{k-1}(t_0) = 0$, we use the equality $f_6(\cdot, 0, 1) = 0$ to conclude $x_{k-1}(t) = 1$ for $t \geq t_0$. Then $k - 1$ is also fixed.

For $\ell = 7$ we have $f_7(\cdot, \cdot, 0) = 1$, $f(\cdot, \cdot, 1) = 0$. Boundary condition $x_{n+1}(t) = 0$ for $t \geq 0$ implies the result. ∎

Denote by $\{T^{(\ell)}\}$ the set of periods taken by the orbits of f_ℓ. We shall use the following elementary device. Assume that f_ℓ, $f_{\ell'}$, are two different rules such that $f_\ell(u_1, u_2, u_3) = f_{\ell'}(u_1, u_2, u_3)$ for each tuple $(u_1, u_2, u_3) \neq (u_1^{(0)}, u_2^{(0)}, u_3^{(0)})$. If we are able to show that the block $(u_1^{(0)}, u_2^{(0)}, u_3^{(0)})$ does never apear in an orbit of $f_{\ell'}$ then we conclude $\{T^{(\ell')}\} \subset \{T^{(\ell)}\}$.

Lemma 4.5. Let $\theta < 0$, $d < 0 < a$. Then the orbits of the Bounded Neural Network $\mathcal{N} = (I, a, c, d, \theta)$ associated to f_8, f_9, f_{10}, f_{11} are of period $T \leq 2$.

Proof. First let us prove $\{T^{(8)}\} \subset \{T^{(9)}\}$. It suffices to show that the block 101 does never appear in an orbit of f_8. Remark that the state '0' can come from the block 001 or from a tuple of the form $(\cdot, 1, \cdot)$. In the first case we cannot produce the rightmost '1' of (101) and in the second we arrive at a contradiction with the boundary condition.

Now we show the lemma for f_9, f_{10}. Let k be the leftmost coordinate ≥ 1 such that $x_{k'}(t)$ is of period ≤ 2 for any $k' \geq k$. We shall prove that necessarily $k = 1$. Assume the contrary, we shall arrive to a contradiction.

Suppose $x_k(t) = 0$ for any t. Both rules f_9, f_{10} satisfy $\mathrm{sign}(c - \theta) = \mathrm{sign}(a + c - \theta)$ so $x_{k-1}(t)$ is of period-2.

Assume $x_k(2t) = 0$, $x_k(2t + 1) = 1$ for any t. If $x_{k-1}(2t_0) = 0$ for some t_0 we deduce $x_{k-1}(2t + 1) = 1$, $x_{k-1}(2t) = 0$ for any t. Now suppose $x_{k-1}(2t_0) = 1$.

If $\ell = 9$ we get $x_{k-1}(2t_0 + 1) = 0$, $x_{k-1}(2t_0 + 2) = 0$. From the last condition we obtain $x_{k-1}(2t) = 0$, $x_{k-1}(2t + 1) = 1$ for any t.

$$
\begin{array}{ccc}
1 & 0 & 2t_0 \\
0 & 1 & 2t_0 + 1 \\
0 & 0 & 2t_0 + 2 \\
1 & 1 &
\end{array}
$$

If $\ell = 10$ we get $x_{k-1}(2t_0 + 1) = 1$, so $x_{k-1}(2t) = 0$ and $x_{k-1}(2t + 1) = 1$ for any t. In both cases $x_{k-1}(t)$ is of period-2.

$$
\begin{array}{ccc}
1 & 0 & 2t_0 \\
1 & 1 & 2t_0 + 1 \\
0 & 0 & \\
1 & 1 &
\end{array}
$$

In both cases $x_{k-1}(t)$ is of period-2.

Finally assume that $x_k(t) = 1$ is fixed for any t. It is easy to show for f_9 that the coordinate $x_{k-1}(t) = 0$ is fixed for any t. For f_{10} the coordinate $k - 1$ is of period 2: $x_{k-1}(2t) = 0$, $x_{k-1}(2t + 1) = 1$ for any t. We conclude $T \leq 2$ for $\ell = 9, 10$.

Finally let us prove the result for f_{11}. Take $\{x(t)\}$ an orbit of f_{11}. Note by k the rightmost coordinate $\leq n$ such that $x_k(t)$ is of period ≤ 2. If $k < n$ we shall arrive at a contradiction.

If $x_k(t) = 0$ for any t we use $f_{11}(0, 1, \cdot) = 0$, $f_{11}(0, 0, \cdot) = 1$ to deduce $x_{k+1}(t)$ is of period 2. From $f_{11}(1, 0, \cdot) = 1$, $f(\cdot, 1, 1) = 0$ we find that condition $x_k(t) = 1$ for any t is impossible. So assume $x_k(2t) = 0$, $x_k(2t + 1) = 1$. If $x_{k+1}(2t_0) = 1$ for some t_0 we deduce $x_{k+1}(2t) = 1$, $x_{k+1}(2t + 1) = 0$. Then suppose $x_{k+1}(2t) = 0$ for any t, which necessarily implies $x_{k+1}(2t + 1) = 1$, $x_{k+2}(2t + 1) = 1$. Then $x_{k+1}(t)$ is of period ≤ 2. Hence the lemma is shown. ∎

Lemma 4.6. Let $\theta < 0$, $d < 0 < a$. Then the orbits of the Bounded Neural Network $N = (I, a, c, d, \theta)$ associated to f_{12} are of period $T = 4$.

Proof. Pick $\{x(t)\}$ an orbit of f_{12}. Since $f_{12}(\cdot,\cdot,0) = 1$ we deduce $x_n(t) = 1$ is fixed for any t. We shall examine the orbit $x'(t) = (x_i'(t) : i \in I' - \{n\})$. Obviously the periods of $x(t)$ and $x'(t)$ are the same.

Define $A' = (a_{ij}' : i,j \in I')$ by $a_{ii}' = 0$ and $a_{ij}' = |\frac{d}{a}|^i a_{ij}$ if $i \neq j$ (the coefficients a_{ij} are those defined in (4.18)). Take $b' = (b_i' : i \in I')$ such that $b_i' = |\frac{d}{a}|^i \theta$ if $i \leq n - 2$ and $b_{n-1}' = |\frac{d}{a}|^{n-1}(\theta - d)$. By taking into account the boundary condition $x_n(t) = 1$ and the fact that f_{12} does not depend on the central cell, we get:

$$x'(t+1) = \bar{\mathbb{1}}(A'x'(t) - b') \quad \text{for } t \geq 1$$

By construction the matrix A' is antisymmetric.

Now note that the restrictions we impose on coefficient a are: $a > \theta$, $a > -d + \theta$, $a > -d + (\theta - c)$. Since $\theta < 0$, $(\theta - c) < 0$ we can assume that the inequality $a < -d$ holds.

Since all inequalities are strict we can assume that every $(u_1, u_2) \in \{0,1\}^2$ satisfies: $au_1 + du_2 - \frac{a+d}{2} \neq 0$, $du_2 - \frac{d}{2} \neq 0$, $au_1 + d - \frac{a}{2} \neq 0$.

We shall prove that:

$$\bar{\mathbb{1}}(A'x'(t) - b') = \bar{\mathbb{1}}(A'x'(t) - b'') \quad \text{where } b_i'' = \frac{1}{2}\sum_{j \in I'} a_{ij} \quad \text{for } i \in I' \quad (4.21)$$

By Proposition 3.14 for antisymmetric iterations and from the fact that for our matrix A' condition (3.33) will be verified, we shall be able to conclude that the evolution $x'(t+1) = \mathbb{1}(A'x'(t) - b') = \mathbb{1}(A'x'(t) - b'')$ is of period $T = 4$.

Property (4.21) for coordinate $i = 1$ is equivalent to

$$\mathbb{1}(bx_2'(t) - \theta) = \mathbb{1}(dx_2'(t) - \frac{d}{2}).$$

This equality is satisfied in both cases $x_2'(t) = 0$ or 1, because $d < 0$ and $d - \theta < 0$.

For coordinates $1 < i < n - 1$ equality (4.21) is written as:

$$\mathbb{1}(ax_{i-1}'(t) + dx_{i+1}' - \theta) = \mathbb{1}(ax_{i-1}'(t) + dx_{i+1}'(t) - \frac{a+d}{2})$$

The assumptions $a + d - \theta > 0$, $a + d < 0$ implies this result for any pair $(x_{i-1}'(t), x_{i+1}'(t)) \in \{0,1\}^2$.

For $i = n - 1$ we must prove: $\mathbb{1}(ax_{n-2}'(t) + d - \theta) = \mathbb{1}(ax_{n-2}'(t) + d - \frac{a}{2})$. This follows from $\frac{a}{2} + d < 0$. Then the result. ■

Lemma 4.7. Let $\theta < 0$, $d < 0 < a$. Then the orbits of the Bounded Neural Network $\mathcal{N} = (I, a, c, d, \theta)$ associated to the local functions f_{13}, f_{14}, f_{15}, f_{16}, f_{17} are of period $T = 3$.

Proof. First, let us prove that any orbit $\{x(t)\}$ of f_{13} has period $T = 3$. As $x_k(t) = 1$, $k < n$ implies $x_{k+1}(t+1) = 1$ we deduce that there must exist some t_0 for which $x_1(t_0) = 0$. Hence $x_1(t_0 - 1) = x_2(t_0 - 1) = 1$ and $x_2(t_0) = x_3(t_0) = 1$. Then $x_1(t_0 + 1) = x_3(t_0 + 1) = 1$, $x_2(t_0 + 1) = 0$ and $x_1(t_0 + 2) = x_2(t_0 + 2) = 1$. We conclude $x_1(t_0 + 3) = 0$ then the coordinates $x_1(t)$, $x_2(t)$ are of period 3.

$k = 0$	1	2	
0	1	1	$t_0 - 1$
0	0	1	1	t_0
0	1	0	1	$t_0 + 1$
0	1	1	$t_0 + 2$
0	0	1	$t_0 + 3$

Let k be the rightmost coordinate $\leq n$ for which $x_{k'}(t)$ is of period 1 or 3 for any $k' \leq k$. Let us show $k = n$. If $x_k(t) = 0$ for any t, by the above analysis we deduce that $x_{k+1}(t)$ is also of period 3. The condition $x_k(t) = 1$ for any t, implies $x_{k+1}(t) = 1$ for any t. On the other hand the case $x_k(t) = 0$, $x_k(t+1) = 0$ is impossible, so the only case we must study is $x_k(3t) = 0$, $x_k(3t + 1) = 1$, $x_k(3t + 2) = 1$ for any t. We deduce $x_{k+1}(3t + 2) = 1$, $x_{k+1}(3t) = 1$, $x_{k+2}(3t) = 1$. We conclude $x_{k+1}(3t + 1) = 0$ then $x_{k+1}(t)$ is also of period 3.

k	$k + 1$	$k + 2$	
1	1		$3t - 1$
0	1	1	$3t$
1	0		$3t + 1$
1	1		$3t + 2$

Now we show that any orbit $\{x(t)\}$ of f_{14} has period $T = 3$. Since $f_{14}(\cdot, \cdot, 0) = 1$ the coordinate $x_n(t) = 1$ is fixed. Condition $f_{14}(\cdot, 1, 1) = 0$ implies that there exists t_0 such that $x_{n-1}(t_0) = 0$. On the other hand if $x_{n-1}(t) = 0$ is fixed we arrive at a contradiction when we examine the coordinate $x_{n-2}(t)$ (the condition $x_{n-2}(t) = 0$ implies $x_{n-2}(t+1) = 1$, which contradicts $x_{n-1}(t+2) = 0$). Then we can assume $x_{n-1}(t_0 + 1) = 1$. This implies $x_{n-2}(t_0) = 1$, $x_{n-1}(t_0 + 2) = 0$, $x_{n-2}(t_0 + 1) = 1$, $x_{n-2}(t_0 + 2) = 0$, $x_{n-1}(t_0 + 3) = 0$, $x_{n-2}(t_0 + 3) = 1$. Then the block $(x_{n-2}(t), x_{n-1}(t), x_n(t))$ is of period 3.

$n - 2$	$n - 1$	n	$n + 1$	
1	0	1	0	t_0
1	1	1	0	$t_0 + 1$
0	0	1	0	$t_0 + 2$
1	0	1	0	$t_0 + 3$

Let k be the leftmost coordinate ≥ 1 such that $x_{k'}(t)$ is of period $T = 1$ or 3 for $k' \geq k$. We shall prove $k = 1$. In fact if this is not the case and $x_k(t) = 0$ for any t or $x_k(t) = 1$ for any t we arrive at a contradiction because $x_{k-1}(t)$ is also of period 1 or 3.

Assume $x_k(3t) = 0$, $x_k(3t + 1) = 0$, $x_k(3t + 2) = 1$ for any t. We deduce $x_{k-1}(3t + 1) = 1$, $x_{k-1}(3t + 2) = 1$. So $x_{k-1}(3t) = 0$ and $x_{k-1}(t)$ is of period 3.

Assume $x_k(3t) = 1$, $x_k(3t + 1) = 1$, $x_k(3t + 2) = 0$. So $x_{k-1}(3t) = 1$, $x_{k-1}(3t + 1) = 0$ and $x_{k-2}(3t + 2) = 1$. If for some t_0 we have $x_{k-1}(3t_0 + 2) = 0$ we easily deduce that $x_{k-1}(3t+2) = 0$ for any t. The other case is $x_{k-1}(3t+2) = 1$ for any t, and in both cases $x_{k-1}(t)$ os of period 3. Hence the result for f_{14},

Now let us prove $\{T^{(15)}\} \subset \{T^{(17)}\}$. It suffices to show that the block 101 cannot appear in an orbit of f_{15}. If the rightmost '1' came from 000 it would be impossible to produce '0', if it came from a tuple of the form $(1, \cdot, \cdot)$ the leftmost '1' could not appear.

In an analogous form we shall prove $\{T^{(16)}\} \subset \{T^{(17)}\}$. We must show that 111 does not appear in an orbit of f_{16}. It suffices to examine the existence of a block 11. The leftmost '1' can arise from 000 or $(1, \cdot, \cdot)$. It is impossible that it be produced by 101 because if it were so the second '1' of the block 111 would not appear. If it came from 100 we would necessarily deduce that the block is 1000. But $x_k(t - 1) = 0$ for $k \neq n + 1$ implies $x_{k-1}(t - 2) = 0$ and we would also arrive at a contradiction.

$$
\begin{array}{cccc}
0 & 0 & 0 & \\
0 & 0 & 0 & t - 2 \\
1 & 0 & 0 & 0 \quad t - 1 \\
& 1 & 1 & t
\end{array}
$$

Then the leftmost '1' of the block 11 comes from either $(1, 1, \cdot)$ or 000. Denote by k the first coordinate which verifies $x_k(t) = x_{k+1}(t) = x_{k+2}(t) = 0$ and $x_{k+1}(t + 1) = x_{k+2}(t + 1) = 1$. We deduce $x_{k+3}(t) = 0$. Then $x_k(t - 1) = x_{k+1}(t - 1) = x_{k+2}(t - 1) = 0$, so we arrive at a contradiction with $x_{k+1}(t) = 0$. Hence $\{T^{(16)}\} \subset \{T^{(17)}\}$.

$$
\begin{array}{cccc}
k & k+1 & k+2 & k+3 \\
0 & 0 & 0 & t - 1 \\
0 & 0 & 0 & 0 \quad t \\
1 & 1 & & t + 1
\end{array}
$$

The lemma will be shown when we prove that each orbit of f_{17} is of period $T^{(17)} = 3$. Let $\{x(t)\}$ be an orbit of f_{17}. First let us show that up to a mod(3) integer the coordinate $x_n(t)$ is of the form:

$$x_n(3t) = 1, \quad x_n(3t + 1) = 1, \quad x_n(3t + 2 = 0 \quad \text{for any } t \tag{4.22}$$

From the boundary condition $x_{n+1}(t) = 0$ and $f_{17}(\cdot, 0, 0) = 1$ we deduce that there exists some t such that $x_n(t) = 1$. On the other hand $f_{17}(0, 1, 0) = 0$ and

$f_{17}(\cdot, 1, 1) = 0$ imply that there exists some '0' in coordinate n, so we may assume for some t_0: $x_n(t_0) = 1$, $x_n(t_0 + 1) = 0$, $x_n(t_0 + 2) = 1$. Let us prove $x_n(t_0 + 3) = 1$.

Denote by $s_0 = \inf\{s \geq 1 : x_{n-s}(t_0 - s) = 0\}$. Obviously such s_0 exists because of the boundary condition $x_0(t) = 0$. If $s_0 = 1$, i.e. $x_{n-1}(t_0 - 1) = 0$, we get $x_n(t_0 - 1) = 0$, $x_{n-1}(t_0) = 1$, which contradicts $x_n(t_0 + 1) = 0$. So $s_0 > 1$. We have $x_{n-s_0}(t_0 - s_0) = 0$, $x_{n-s_0}(t_0 - s_0) = 0$, $x_{n-s_0+2}(t_0 - s_0) = 0$. See Figure 4.7.

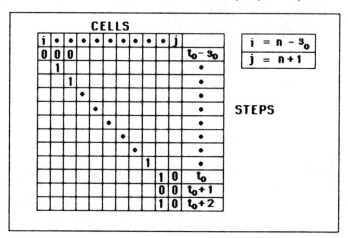

Figure 4.7. Period 3, analysis 1.

Then $x_{n-s_0}(t_0 - s_0 + 1) = 1$, $x_{n-s_0}(t_0 - s_0 + 2) = 0$. Since $x_{n-s_0+2}(t_0 - s_0 + 2) = 1$ we deduce $x_{n-s_0+1}(t_0 - s_0 + 3) = 0$, so $x_{n-s_0+s}(t_0 - s_0 + s + 2) = 0$ for $0 \leq s \leq s_0 - 2$. We have $x_{n-1}(t_0) = 0$, so $x_{n-1}(t_0 + 1) = 0$ and $x_{n-2}(t_0 + 1) = 0$. We get $x_{n-1}(t_0 + 2) = 1$, then $x_n(t_0 + 3) = 0$. We also obtain $x_{n-1}(t_0 + 3) = 0$, so we conclude $x_n(t_0 + 4) = 0$, $x_n(t_0 + 5)$ (see Figure 4.8). Hence $x_n(t)$ has the form given by (4.22). See Figure 4.8.

Now let $k \geq 1$ be the smallest coordinate such that $x_{k'}(t)$ is of period 3 for any $k \leq k' \leq n$. Suppose $k > 1$. By hypothesis $x_k(t)$ can only take two forms. First assume that it satisfies:

$$x_k(3t) = 1, \quad x_k(3t + 1) = 1, \quad x_k(3t + 2) = 0 \quad \text{for any } t \qquad (4.23)$$

We deduce $x_{k-1}(3t) = 1$, $x_{k-1}(3t + 1) = 0$. If $x_{k-2}(3t + 2) = 0$ we obtain $x_{k-1}(3t + 2) = 0$, so $x_{k-2}(3t + 1) = 0$, which contradicts $x_{k-2}(3t + 2) = 0$. Then $x_{k-2}(3t + 2) = 1$ for any t.

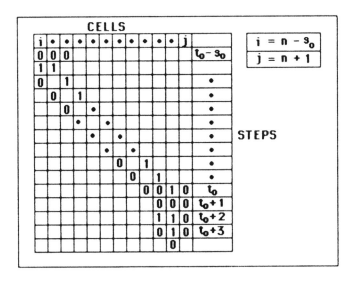

Figure 4.8. Period 3, analysis 2.

Pick t_0. We shall prove $x_{k-1}(3t_0 + 2) = 0$ which shows coordinate $k - 1$ is also of period 3. From the boundary condition $x_0(t) = 0$ for any t, the number $s_0 = \inf\{s \geq 0 : x_{k-1-s}(3t_0 - s) = 0\}$ does exist. Put $k_0 = k - s_0$; the situation is represented in Figure 4.9.

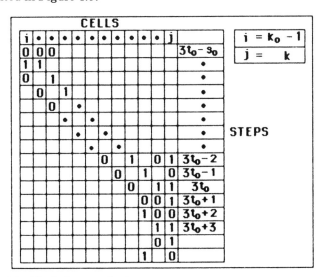

Figure 4.9. Period 3, analysis 3.

Obviously $s_0 \geq 2$. If $s_0 = 2$ we deduce $x_{k-3}(3t_0 - 2) = 0$, $x_{k-2}(3t_0 - 2) = 0$ so $x_{k-1}(3t_0 - 1) = 0$, $x_{k-3}(3t_0 - 1) = 1$. Then $x_{k-3}(3t_0) = 0$, $x_{k-2}(3t_0 + 1) = 0$. We conclude $x_{k-1}(3t_0 + 2) = 0$.

Take $s_0 > 2$. We have: $x_{k-1-s_0}(3t_0 - s_0) = x_{k-s_0}(3t_0 - s_0) = x_{k+1-s+_0}(3t_0 - s_0) = 0$, $x_{k-1-(s_0-1)}(3t_0 - (s_0 - 1)) = 1$. We deduce $x_{k-1-s_0}(3t_0 - (s_0 - 1)) = 1$, $x_{k-1-s_0}(3t_0 - (s_0 - 2)) = 0$. From the definition of s_0 we have $x_{k-1-s}(3t_0 - s) = 1$ for any $0 \leq s < s_0$. So we obtain:

$$x_{k-1-s}(3t_0 - (s - 2)) = 0 \quad \text{for any } 1 \leq s < s_0.$$

Write the case $s = 1$, i.e. $x_{k-2}(3t_0 + 1) = 0$. We get $x_{k-1}(3t_0 + 2) = 0$. So the result holds, i.e. if $x_k(t)$ takes the form (4.23) the coordinate $k - 1$ is also of period 3.

Now assume that $x_k(t)$ satisfies:

$$x_k(3t) = 0, \ x_k(3t + 1) = 0, \ x_k(3t + 2) = 1 \quad \text{for any } t \tag{4.24}$$

We must necessarily have $x_{k-1}(3t) = 0$, $x_{k-1}(3t + 1) = 1$. So if $x_{k-1}(3t + 2) = 1$ for any t we have shown that $x_{k-1}(t)$ is of period $T = 3$. Then assume that some t_0 satisfies $x_{k-1}(3t_0 + 2) = 0$. We deduce that $x_{k-2}(3t_0 + 2) = 0$, $x_{k-2}(3t_0 + 1) = 0$, $x_{k-3}(3t_0 + 1) = 0$, $x_{k-3}(3t_0 + 2) = 1$, $x_{k-2}(3t_0) = 1$, $x_{k-3}(3t_0) = 0$. Also $x_{k-4}(3t_0) = 0$, $x_{k-4}(3t_0 + 1) = 1$.

$k-4$	$k-3$	$k-2$	$k-1$	k	
				1	$3t_0 - 1$
0	0	1	0	0	$3t_0$
1	0	0	1	0	$3t_0 + 1$
	1	0	0	1	$3t_0 + 2$
			0	0	

Now assume $x_{k-4}(3t_0 - 1) = 1$. We get $x_{k-3}(3t_0 - 1) = 1$, $x_{k-2}(3t_0 - 1) = 1$. Now $x_{k-1}(3t_0 - 1) = 0$ contradicts $x_{k-1}(3t_0) = 0$ and $x_{k-1}(3t_0 - 1) = 1$ contradicts $x_{k-2}(3t_0) = 1$. Then $x_{k-4}(3t_0 - 1) = 0$.

Assume that there exists some $s \geq 1$ such that $x_{k-4-s}(3t_0 - 1 - s) = 1$. Call $s_0 = \inf\{s \geq 1 : x_{k-4-s}(3t_0 - 1 - s) = 1\}$. We deduce $x_{k-4-s_0}(3t_0 - 1 - s_0) = x_{k-3-s_0}(3t_0 - 1 - s_0) = x_{k-2-s_0}(3t_0 - 1 - s_0) = 1$ and $x_{k-4-s}(3t_0 - 1 - s) = 0$ for $0 \leq s < s_0$. Put $k_0 = k - 4 - s_0$; then the situation is as indicated in Figure 4.10.

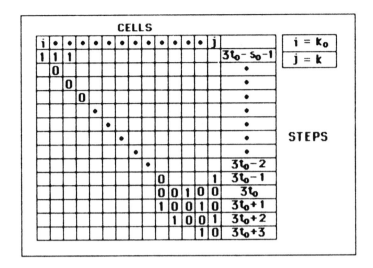

Figure 4.10. Period 3, analysis 4.

From $x_{k-4}(3t_0) = 0$, $x_{k-4}(3t_0 - 1) = 0$ we deduce $x_{k-5}(3t_0 - 1) = 0$, $x_{k-3}(3t_0 - 1) = 1$. Hence

$$x_{k-5-s}(3t_0 - 1 - s) = 0, \quad x_{k-3-s}(3t_0 - 1 - s) = 1 \quad \text{for any } 0 \le s \le s_0 - 1.$$

Then $x_{k-1-s_0}(3t_0 - 1 - s_0) = 0$, $x_{k-1-s_0}(3t_0 - s_0) = 1$. From the equality $x_{k-1-s_0}(3t_0 - s_0 + 1) = 1$ we deduce $x_{k-4-s_0}(3t_0 - s_0) = 0$ and $x_{k-4-s_0}(3t_0 - s_0 + 1) = 1$. By using the same idea we get:

$$x_{k-3-s}(3t_0 - 1 - s) = 1, \quad x_{k-2-s}(3t_0 - 1 - s) = 1, \quad x_{k-1-s}(3t_0 - 1 - s) = 0$$
for any $0 \le s \le s_0$.

Put $t_1 = 3t_0 - s_0$; the situation is as indicated in Figure 4.11.

Evaluating the last quantity at $s = 0$ we get: $x_{k-1}(3t_0 - 1) = 0$. So there is a contradiction because $x_{k-2}(3t_0 - 1) = x_k(3t_0 - 1) = 1$ and $x_{k-1}(3t_0) = 1$.

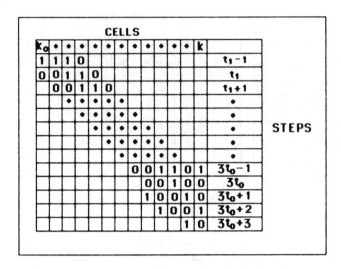

Figure 4.11. Period 3, analysis 5.

CELLS

0	1	2	3	•	•	•	•	•	•	•	•	•	k	STEPS
0														t_2
0	0													t_2+1
0	0	0	1											t_2+2
0	1	0	0	1										t_2+3
0		1	0	0	•									t_2+4
0		1	0	•	•									•
0		1	•	•	•									•
0			•	•	•	•								•
0				•	•	•	1							$3t_0-2$
0				•	•	0	1					1		$3t_0-1$
0					•	0	0	1	0	0				$3t_0$
0						1	0	0	1	0				$3t_0+1$
0							1	0	0	1				$3t_0+2$
0								1	0	0				$3t_0+3$
0									1	0				$3t_0+4$
0										1				$3t_0+5$

Figure 4.12. Period 3, analysis 6.

Then for any $s \geq 1$ we must have $x_{k-4-s}(3t_0 - 1 - s) = 0$. We can also write $x_{k-1-s}(3t_0 + 2 - s) = 0$ for any $0 \leq s \leq k - 1$. Let us examine what happens at coordinate 1. Recall that the diagonal $(k - 1 - s, 3t_0 + 2 - s)$ touches coordinate 0 at time $3t_0 + 3 - k$. By the above analysis we also have $x_{k-1-s}(3t_0 + 3 - s) = 0$ and $x_{k-1-s}(3t_0 + 4 - s) = 0$ for any $0 \leq s \leq k - 1$. Put $t_2 = 3t_0 + 3 - k$; the situation as as indicated in Figure 4.12.

Since $x_0(3t_0 + 6 - k) = 0$, $x_1(3t_0 + 6 - k) = 1$, $x_2(3t_0 + 6 - k) = 0$ we get $x_1(3t_0 + 7 - k) = 0$. Combining this last equality with $x_2(3t_0 + 7 - k) = 1$, $x_3(3t_0 + 7 - k) = 0$ we obtain $x_2(3t_0 + 8 - k) = 0$. In general we deduce:

$$x_s(3t_0 + 6 + s - k) = 0 \quad \text{for any } 0 \leq s \leq k - 1$$

In particular $x_{k-1}(3t_0 + 5) = 0$. Hence $x_{k-1}(3t_0 + 2) = 0$ implies $x_{k-1}(3t_0 + 5) = 0$. So $x_{k-1}(3t_0 + 2) = 0$ for any t. The result is proved. Furthermore in this last case we have shown that for any $1 \leq k' \leq k$ the structure of $x_{k'}(t)$ is $x_{k'}(3t - (k - k')) = 0$, $x_{k'}(3t + 1 - (k - k')) = 0$, $x_{k'}(3t + 2 - (k - k')) = 1$. ∎

Theorem 4.3. [Sh2] Consider the one-dimensional Bounded Neural Network $\mathcal{N} = (I = \{1, ..., n\}, a, c, d, \theta)$ with fixed boundary conditions $\mathcal{B}_0 : x_0(t) = x_{n+1}(t) = 0$ for any t (respectively $\mathcal{B}_1 : x_0(t) = x_{n+1}(t) = 1$ for any t). Then its orbits are of period $T \leq 4$.

Proof. For boundary conditions $\mathcal{B}_0$ the result follows from lemmas 4.1 through 4.7. On the other hand, suppose that the local function f with boundary condition $\mathcal{B}_1$. Associate to it the following local function g_f defined by $g_f(u_1, u_2, u_3) = 1 - f(1 - u_1, 1 - u_2, 1 - u_3)$, and iterate g_f with boundary condition $\mathcal{B}_0$. It is easy to see that the orbits of f and g_f are in one-to-one correspondance, hence the overall result. ∎

Remarks. Consider the one-dimensional Bounded Neural Network $\mathcal{N} = (I, a, c, d, \theta)$ with boundary conditions $\mathcal{B}_0$. Its associated local function is denoted f. Then:
1. If it does not posses non-spontaneous excitation (so $\theta > 0$), we have $T \leq 2$ (this follows from Lemma 4.2).
2. If its period is $T = 3$ then its local function depends on the central cell i.e. there exists $(u_1, u_2, u_3) \in \{0,1\}^3$ such that $f(u_1, u_2, u_3) \neq f(u_1, 1 - u_2, u_3)$. This assertion can be verified in Table 4.1 of signs and periods.
3. Assume that f does not depend on the central cell i.e. $f(u_1, u_2, u_3) = f(u_1, 1 - u_2, u_3)$ for each $(u_1, u_2, u_3) \in \{0,1\}^3$.
    Then its period satisfies $T = 1$ or 4. Let us be more precise.

If $\theta > 0$ any such f posseses period $T = 1$ (Lemma 4.1).

Let $\theta < 0$, $d < 0 < a$. In Table 4.1 of signs and periods we can find all the cases in which f does not depend on the central cell. We must verify:

$$\text{sign}(c - \theta) = \text{sign}(-\theta) = 1, \quad \text{sign}(a + c - \theta) = \text{sign}(a - \theta) = 1,$$
$$\text{sign}(c + d - \theta) = \text{sign}(d - \theta), \quad \text{sign}(a + c + d - \theta) = \text{sign}(a + c - \theta).$$

They are:

Case	Period
f_1	$T^{(1)} = 1$
f_7	$T^{(7)} = 1$
f_{11}	$T^{(11)} = 4$

4. The method we used for proving that the orbits of f_{12} were of period $T = 4$ cannot be applied in the case of the torus, i.e. the fact that our network was open with boundary condition $\mathcal{B}_0$ was essential.

For the torus there exist examples of antisymmetric iterations which do not satisfy condition (3.33) and such that the lengths of their orbits are the same size as the networks. For instance take $I = \mathbb{Z}_n$, the connections $a_{i,i-1} = 1$, $a_{i,i+1} = -1$, $a_{ij} = 0$ for any other pair (i,j) and the thresholds $b_i = \frac{1}{2} \sum_{j \in I} a_{ij} = 0$. It is easy to see that the point $x(0) = (0, 1, ..., 1)$ belongs to an orbit of length n. In this case Proposition 3.14 cannot be applied because condition (3.33) does not hold.

Now vary slighly the framework in order to consider local functions depending on the left and the right neighbourhood of the central cells.

For any $i \in I = \{1, ..., n\}$ we take the neighbourhood $\bar{V}_i = \{i - 1, i + 1\}$. Let $Q = \{0, 1\}$ be the state set and $f : Q^2 \to Q$ be a local function. We impose fixed boundary conditions, for instance $\mathcal{B}_0 : x_0(t) = x_{n+1}(t) = 0$ for any t. Then the evolution of the automaton $\mathcal{A} = (I, \bar{V}_i : i \in I), \{0, 1\}, f)$ is:

$$x_i(t + 1) = f(x_{i-1}(t), x_{i+1}(t)) \text{ for each } i \in I.$$

We wonder what kind of such f's can be realized by a Bounded Neural Network $\mathcal{N} = (I, a, 0, d, \theta)$. Develop the equalities so that this occurs. We must have:

$$\theta \leq 0 \text{ if } f(0,0) = 1 \qquad a \geq \theta \text{ if } f(1,0) = 1$$
$$\theta > 0 \text{ if } f(0,0) = 0 \qquad a < \theta \text{ if } f(1,0) = 0$$

$$d \geq \theta \text{ if } f(0,1) = 1 \qquad a + d \geq \theta \text{ if } f(1,1) = 1$$
$$d < \theta \text{ if } f(0,1) = 0 \qquad a + d < \theta \text{ if } f(1,1) = 0$$

Note that $\theta > 0$, $a \geq \theta$, $d \geq \theta$ implies $a + d \geq \theta$. Then the XOR rule given by $f(0,0) = f(1,1) = 0$, $f(1,0) = f(0,1) = 1$ cannnot be realized with some $\mathcal{N} = (I, a, 0, d, \theta)$.

We also observe that $\theta \leq 0$, $a \leq \theta$, $d \leq \theta$ implies $a + d \leq \theta$. Then the $\overline{XOR}$ rule given by $f(0,0) = f(1,1) = 1$, $f(1,0) = f(0,1) = 0$ cannot be obtained with some $\mathcal{N} = (I, a, 0, d, \theta)$.

We must point out that the XOR and the $\overline{XOR}$ rules belong to class 3 of Wolfram classification [W2].

It is easily shown (by developping all the cases if necessary) that any other local rule f can be realized with a Neural Network of the type $\mathcal{N} = (I, a, 0, d, \theta)$. Hence:

Corollary 4.1. Consider the finite automaton $\mathcal{A} = (I, (\bar{V_i} : i \in I), \{0,1\}, f)$ with fixed boundary condition (we can take $\mathcal{B}_0$ or $\mathcal{B}_1$). Then if f is not the XOR or the $\overline{XOR}$ rule the orbits of the automaton $\mathcal{A}$ are of period $T \leq 4$.

Proof. From Theorem 4.2 and the above discussion. ■

Remark. This corollary asserts that bounded automata with $Q = \{0,1\}$ and next-nearest interactions have a simple dynamics: they belong to the Wolfram's class 2 except the XOR and the $\overline{XOR}$ whose belong to class 3.

In Figures 4.13, 4.14 and 4.15 we give some typical patterns generated by the one-dimensional Bounded Neural Network of Figure 4.6 with the local rules in Table 4.1. All the patterns are generated by the same initial condition on the set of cells $I = \{1, ..., 250\}$. While we have proved that, in all the cases, the steady state is simple ($T \leq 4$), some complexity is observed in the transient phase. In this context see the Bounded Neural Network associated to f_{11}, f_{15} and f_{17}. For f_{11} the dynamics takes a long time to reach the two-period behaviour and the frontier between the two-cycle pattern generated on the left and the pattern on the right has a complex interaction. Similar phenomena are observed for f_{15} and f_{17} but on 3-periodic patterns.

For the other rules the dynamics is simpler: fast convergence to steady state and in most of the cases the initial configuration is shifted from one extreme to the other and one, two or three cycles are imposed.

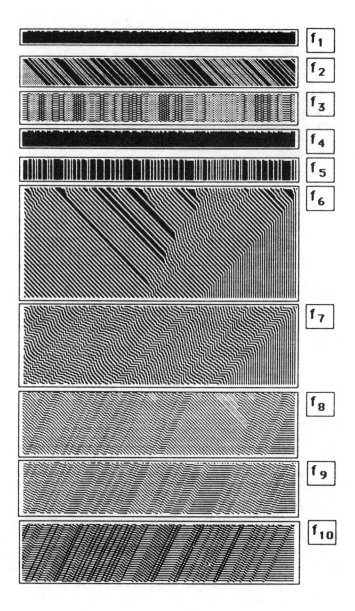

Figure 4.13. Dynamics of the one-dimensional BNN associated to the local rules f_i, $i = 1, ..., 10$ of table 4.1 on the cellular space $I = \{1, ..., 250\}$.

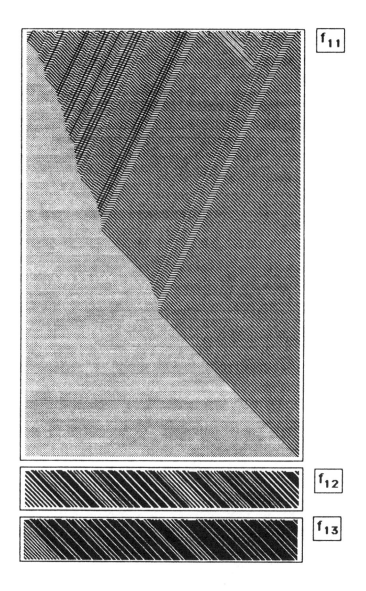

Figure 4.14. Dynamics of the one-dimensional BNN associated to the local rules f_{11}, f_{12}, f_{13} for the cellular space $I = \{1, ..., 250\}$.

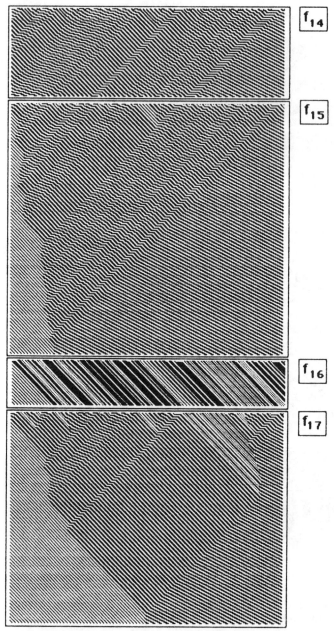

Figure 4.15. Dynamics of the one-dimensional BNN associated to the local rules f_{14}, f_{15}, f_{16} f_{17} for the cellular space $I = \{1, ..., 250\}$.

4.5. Two-Dimensional Bounded Neural Networks

Let $I_1 = \{1, ..., n_1\}$, $I_2 = \{1, ..., n_2\}$. The set of cells of the network is $I = I_1 \times I_2$. The neighbourhood V_i of any $i = (i_1, i_2) \in I$ is $V_i = V_i^N \cap I$ where V_i^N is the von-Neumann neighbourhood in $\mathbb{Z}^2$. The interaction matrix $A = (a_{ij} : i, j \in I)$ is uniform:

$$a_{(i_1,i_2),(i_1-1,i_2)} = a_1, \quad a_{(i_1,i_2),(i_1+1,i_2)} = d_1, \quad a_{i,i} = c$$

$$a_{(i_1,i_2),(i_1,i_2-1)} = a_2, \quad a_{(i_1,i_2),(i_1,i_2+1)} = d_2$$

The threshold $b = \bar{\theta}$ is the same for every point: $b_i = \theta$ for each $i \in I$.

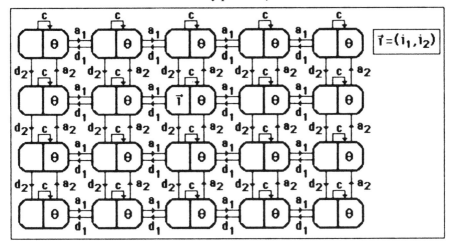

Figure 4.16. Two-dimensional BNN, $n_1 \times n_2 = 4 \times 5$.

The state set is $Q = \{0, 1\}$ so the configuration space is $C = \{0, 1\}^{|I|}$. The evolution of this network $N = (I, A, b)$ is given by the equation:

$$x_i(t+1) = \mathbb{1}\left(\sum_{j \in V_i} a_{ij} x_j(t) - \theta\right) \quad i \in I$$

If we impose the boundary condition:

for $i_1 \in \{0, n_1 + 1\}$, $i_2 \in \{0, n_2 + 1\} : x_{(i_1,i_2)}(t) = 0$ for any $t \geq 0$

This automaton will be called a two dimensional Bounded Neural Network.

The above evolution can be written in the form:

$$x_i(t+1) = f(x_j(i) : j \in V_i^N) \quad \text{for any } i \in I \text{ with}$$

$$f(x_j : j \in V_i^N) = \mathbb{1}\Big(\sum_{j \in V_i^N} a_{ij} x_j - \theta\Big)$$

Note that if $a_1 \neq d_1$ or $a_2 \neq d_2$ the matrix of interactions A is not symmetric. As in the one-dimensional case, we have:

Theorem 4.3. Suppose $\text{sign}(a_1) = \text{sign}(d_1)$ and $\text{sign}(a_2) = \text{sign}(d_2)$. Then the two-dimensional Bounded Neural Network $\mathcal{N} = (I, A, \bar{\theta})$ is equivalent to a symmetric Neural Network $\mathcal{N}' = (I, A', b')$ in the following strong sense: there exist a symmetric matrix A' and a vector b' such that:

$$\text{for every } x \in \{0,1\}^{|I|} : \mathbb{1}(Ax - \bar{\theta}) = \mathbb{1}(A'x - b')$$

Therefore the orbits of $\mathcal{N}$ are of period $T \leq 2$.

Proof. If $a_1 = d_1 = 0$ or $a_2 = d_2 = 0$ the situation reduces to the one-dimensional case so Lemma 4.1 implies the result. Hence assume $a_1, d_1, a_2, d_2 \neq 0$.
    Put

$$b'_{(i_1,i_2)} = \Big(\frac{d_1}{a_1}\Big)^{i_1} \Big(\frac{d_2}{a_2}\Big)^{i_2} \theta$$

$$a'_{(i_1,i_2),j} = \Big(\frac{d_1}{a_1}\Big)^{i_1} \Big(\frac{d_2}{a_2}\Big)^{i_2} a_{ij}$$

From the definitions it is obvious that $\mathbb{1}(Ax - \bar{\theta}) = \mathbb{1}(A'x - b')$.
    The proof that A' is symmetric is entirely analogous to the one-dimensional case. In fact a direct computation shows that the following equalities hold:
$a'_{(i_1,i_2),(i_1-1,i_2)} = a'_{(i_1-1,i_2),(i_1,i_2)}, a'_{(i_1,i_2),(i_1+1,i_2)} = a'_{(i_1+1,i_2),(i_1,i_2)},$
$a'_{(i_1,i_2),(i_1,i_2-1)} = a'_{(i_1,i_2-1),(i_1,i_2)}, a'_{(i_1,i_2),(i_1,i_2+1)} = a'_{(i_1,i_2+1),(i_1,i_2)}.$ ∎

Remarks.
1. The particular case $\text{sign}(a_1) = \text{sign}(d_1) = \text{sign}(a_2) = \text{sign}(d_2) = 1$ of the above Corollary was shown in [Sh1].
2. The last Theorem can be easily generalized to the multidimensional case and for any threshold vector b. In fact take the set of cells $I = \prod_{\ell=1}^{m} I_\ell$ with $I_\ell = \{1, ..., n_\ell\}$ and consider the neighbourhoods $V_i = V_i^N \cap I$ where

$$V_i^N = \{j \in I : \sum_{\ell=1}^{m} |j_\ell - i_\ell| \leq 1\}.$$

Put $a_{i,i} = c$, $a_{i,j} = a_\ell$, if $j \in V_i$, $j_\ell = i_\ell - 1$ and $a_{i,j} = d_\ell$, if $j \in V_i$, $j_\ell = i_\ell + 1$. No restrictions are assumed on the vector $b = (b_i : i \in I)$. Then the condition

$$\text{sign}(a_\ell) = \text{sign}(d_\ell) \quad \text{for any } \ell = 1, ..., m$$

implies that the Neural Network $\mathcal{N} = (I, A, b)$ can be symmetrized and its orbits are of period $T \leq 2$.

3. With the results of Chapter 3 we may give bounds on transient lengths of Bounded Neural Networks.

4. If the sign hypothesis is not met, long periods may exist. For instance in the uniform 2-dimensional case take $a_1 = 2$, $d_1 = -1$, $a_2 = -2$, $d_2 = -1$, $c = 0$. Shingai exhibed in [Sh1] non-bounded periods (with the network size). The dynamics of this BNN is exhibited in Figure 4.19 for a random initial configuration. In the steady state a cycle of period $0(n)$ is observed as a rectangle travelling in the cellular space. Other patterns generated by two dimensional BNN are shown in Figures 4.20 and 4.21.

We shall also exhibit a particular family of networks with non-bounded periods which is due to Maass. First let us describe it for $n_1 = 6$, $n_2 = 4$. In this case the initial condition $x(0)$ in Figure 4.17 has a period $T = 10$:

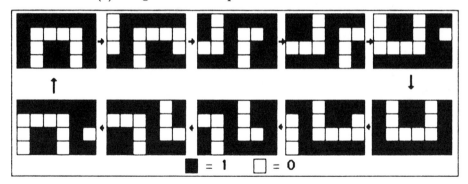

Figure 4.17. Ten-cycle associated to Shingai's rule with coefficients $a_1 = 2$, $d_1 = -1$, $a_2 = -2$, $d_2 = -1$, $x = 0$ in a 4×6 rectangle of cells.

In the general case consider $n_1 = 2m + 2$, $n_2 = 2m$ for some $m \geq 1$. The initial condition $x(0) = (x_{(i_1, i_2)}(0) : i_1 \in I_1, i_2 \in I_2)$ is given by

$$x_{(1, i_2')}(0) = x_{(n_1, i_2')}(0) = x_{(i_1', n_2)}(0) = 1 \quad \text{for } 1 \leq i_1' \leq n_1, \ 1 \leq i_2' \leq n_2$$

$$x_{(2, i_2')}(0) = x_{(n_1 - 1, i_2')}(0) = x_{(i_1', n_2 - 1)}(0) = 0 \quad \text{for } 2 \leq i_1' \leq n_1 - 1, \ i_2' \leq n_2 - 1$$

When $3 \leq i_1 \leq 2m + 1$ we have for any $i_1 \leq i_1' \leq n_1 - i_1 + 1, i_1 - 2 \leq i_2' \leq n_1 - i_1 - 1$:

$$x_{(i_1, i_2')}(0) = x_{(n_1 - i_1 + 1, i_2')}(0) = x_{(i_1', i_1 - 2)}(0) = x_{(i_1', n_1 - i_1 - 1)}(0) = \begin{cases} 1 & \text{if } i_1 \text{ is odd} \\ 0 & \text{if } i_1 \text{ is even} \end{cases}$$

See Figure 4.18:

n_2	1	1	1	1	1	1	1	1	1	1	1	1
	1	0	0	0	0	0	0	0	0	0	0	1
	1	0	1	1	1	1	1	1	1	1	0	1
	1	0	1	0	0	0	0	0	0	1	0	1
m	1	0	1	0	1	y	y	1	0	1	0	1
$m-1$	1	0	1	0	1	y	y	1	0	1	0	1
	1	0	1	1	1	1	1	1	0	1	0	1
	1	0	1	0	0	0	0	0	0	1	0	1
	1	0	1	1	1	1	1	1	1	1	0	1
						$m+1$	$m+2$					n_1

Figure 4.18. $y = 0$ if m is even, $y = 1$ if m is odd.

The size of the network is not arbitrary, it is chosen so that the initial configuration $x(0)$ does not posses a central column of the form $1010...101...0101$. In fact it is easy to show that the existence of a column of this type implies the convergence of $x(0)$ to a fixed point. Then in order to insure a long period we impose the above form of n_1, n_2.

To prove that $x(0)$ belongs to a periodic orbit we must take into account the following facts:

(i) the 1's of $x(0)$ move to the right;

(ii) the local rule verifies:

$$\begin{matrix} & 1 & & & 0 & & & 1 & & & 1 & \\ 0 & & 0 \to 0, & 0 & & 1 \to 0, & 0 & & 1 \to 0, & 0 & 0 & 1 \to 0 \\ & 1 & & & 1 & & & 0 & & & 1 & \end{matrix}$$

therefore we deduce the 0's move to the right;

(iii) the form of the first and second column guarantees the generation of the same configuration but rotated;

(iv) the vertical symmetry of the rule.

By using these facts it can be shown that the period of $x(0)$ is $T = 2(2m+3)$, which is the time that it takes the configuration to complete two rotations.

| ROW 1,2,3 | STEP= 0 1 12 13 | 35 36 51 52 | 60 74 80 90 |
| ROW 4,5,6 | 200 275 280 284 | 105 106 118 119 | 125 126 133 134 |

Figure 4.19. Dynamics of Shingai's rule with $a_1 = 2$, $d_1 = -1$, $a_2 = -2$, $d_2 = -1$, $x = 0$ in a 80×80 rectangle of cells.

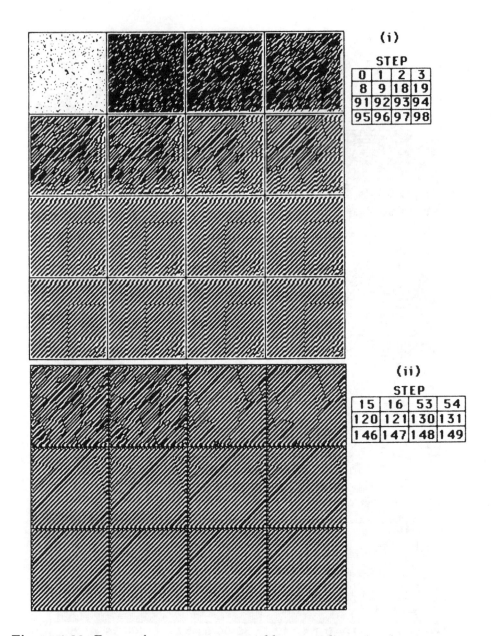

STEP

0	1	2	3
8	9	18	19
91	92	93	94
95	96	97	98

(ii)

STEP

15	16	53	54
120	121	130	131
146	147	148	149

Figure 4.20. Four-cycles patterns generated by a two-dimensional antisymmetric BNN with coefficients $a_1 = a_2 = -1$, $d_1 = d_2 = +1$, $c = 0$, $\theta = 0$. (i) Dynamics on a 75×75 rectangle. (ii) Dynamics on a 75×75 torus.

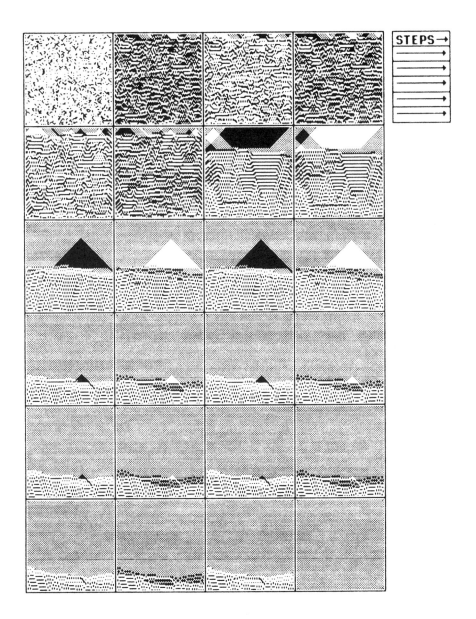

Figure 4.21. Complex convergence to a fixed point of a two-dimensional BNN with coefficients $a_1 = -1$, $a_2 = -1$, $d_1 = +1$, $d_2 = -1$, $c = 0$.

References

[A]   Amari, S., *Homogeneous Nets of Neuron like Elements*, Biol. Cybern., 17, 1975, 211-220.

[B]   Beurle, R.L., *Storage and Manipulation of Information in the Brain*, J. Inst. Elec. Eng., London, 1959, 75-82.

[FC] Farley, B., W.A. Clark, *Activity in Neurons-Like Elements*, in Proc. Fourth London Conference on Information Theory, C. Cherry ed., Butterworths, London, 1961, 242-251.

[GT2] Goles, E., M. Tchuente, *Erasing Multithreshold Automata*, in Dynamical Systems and Cellular Automata, J. Demongeot, E. Goles, M. Tchuente eds., Academic Press, 1985, 47-56.

[GT3] Goles, E., M. Tchuente, *Iterative Behaviour of One-Dimensional Threshold Automata*, Disc. Appl. Maths, 8, 1984, 319-322.

[Ki1] Kitagawa, T., *Cell Space Approaches in Biomathematics*, Math. Bios, 19, 1974, 27-71.

[Ko] Kobuchi, Y., *Signal Propagation in 2-D Threshold Cellular Space*, J. Math. Biol, 3, 1976, 297-312.

[Sh1] Shingai, R., *Maximum Period of 2-Dimensional Uniform Neural Networks*, Information and Control, 41, 1979, 324-341.

[Sh2] Shingai, R., *The Maximum Period Realized in 1-D Uniform Neural Networks*, Trans. IECE, Japan E61, 1978, 804-808.

[T1] Tchuente, M., *Contribution a l'Etude des Methodes de Calcul pour des Systemes de Type Cooperatif*. Thesis, IMAG, Grenoble, France, 1982.

[T3] Tchuente, M., *Evolution de Certains Automatas Cellulaires Uniformes Binaires a Seuil*, Seminaire, 265, IMAG, Grenoble, 1977.

[W2] Wolfram, S., *Universality and Complexity in Cellular Automata*, Physica 10D, 1984, 1-35.

5. CONTINUOUS AND CYCLICALLY MONOTONE NETWORKS

5.1. Introduction

In this chapter we shall extend the results obtained for Neural Networks. This generalization includes continuous state spaces and allows applications in smoothing techniques related to optimization algorithms. Also, this extension turns out to be adequate to study problems arising in statistical physics which will be developed in Chapter 6.

The networks here studied will be in general of the form $\mathcal{G} = (I, A, b, C, f)$ where I is a finite set, A is the connection matrix, b is a vector (in some cases it is null) and C is the set of configurations which is a convex subset of $\mathbb{R}^{|I|}$. The evolution has the shape:

$$F_{\mathcal{G}}(x) = f(Ax - b), \quad \text{where } f : \mathbb{R}^{|I|} \to C.$$

When $C = Q^{|I|}$ for some real subset Q, we use the notation $\mathcal{G} = (I, A, b, Q, f)$.

To approach the continuous state case, three roads are used:

- inspired by the Lyapunov functionals associated to Neural Networks, we are led to consider positive functions which cover the threshold and the sign function while in higher dimensional they include continuous functions. This we do in section 5.2.

- The real continuous case can be studied as a limit of multithreshold functions. This program is fully developed in sections 5.3, 5.4.

- By using convex analysis theory we construct a general Lyapunov functional for networks in which f is a cyclically monotone function. This covers positive and multithreshold networks and constitutes a very large class where steady states are simple (low periods). This task is undertaken in section 5.5. In section 5.7 we generalize this construction.

Besides, dynamical aspects of automata networks can be dealt with by using convex analysis; this approach is developed in [PE,PT1,PT2].

Relations with optimization problems are considered in the framework of positive automata. When the matrix is positive definite the solutions of certain optimization problems appear as fixed points of the automata network dynamics. The reciprocal situation, i.e. when the solutions of some optimization problem can be obtained with automata network evolutions, is a subject in which research has only recently begun (see [An,Ho1,HoT,PA1,PA2]).

In section 5.9 matrices satisfying a less restrictive condition than symmetry are considered. All the dynamical results remain true when symmetry is replaced by this new property.

5.2. Positive Networks

Let $I = \{1, ..., n\}$ be the finite set of sites. Recall that the synchronous evolution of a Neural Network $\mathcal{N} = (I, A, b)$ is given by the equation:

$$x(t+1) = f(Ax(t) - b) \text{ for } x(t) \in C \tag{5.1}$$

where $C = Q^n$ and $Q = \{0, 1\}$. The function $f = \bar{\mathbb{1}} : \mathbb{R}^n \to C$ is the multidimensional threshold function.

In this section we shall deal with generalized networks which are tuples $\mathcal{G} = (I, A, b, C, f)$. In previous chapters the configuration space C was of the form $C = Q^n$, Q being the state space which in all the cases was a real subset, $Q \subset \mathbb{R}$. When $C = Q^n$ we denote $\mathcal{G} = (I, A, b, Q, f)$. But often we must take into account global restrictions, so rather than $C = Q^n$ our set of configurations C is a subset of $\mathbb{R}^n$. The matrix of interactions is $A = (a_{ij} : i, j \in I)$ and $b = (b_i : i \in I)$ is some vector which can vanish. The function $f : \mathbb{R}^n \to C$ sends $\mathbb{R}^n$ into the configuration space and the dynamics of the networks is given by the transformation:

$$F_{\mathcal{G}}(x) = f(Ax - b) \tag{5.2}$$

therefore any trajectory obeys evolution equation (5.1).

In some cases the domain of A and f is a subset $D \subset \mathbb{R}^n$. When this situation occurs we assume that A *preserves* D i.e. $A(D) \subset D$ and $f : D \to C$. In this case we also assume that $D \subset \mathbb{R}^n$ is a convex set, this means:

$$\text{for any} \quad \lambda \in [0, 1], x, x' \in D \text{ implies } \lambda x + (1 - \lambda)x' \in D \tag{5.3}$$

To avoid notational problems we shall not make explicit the dependance of the network $\mathcal{G}$ on the convex subset D preserved by A.

For a particular class of functions f we shall extend the results obtained in previous chapters. In this purpose we are going to examine the hypotheses we must impose on f so that a functional analogous to the one constructed in section 3 can be decreasing for symmetric connections. The existence of Lyapunov functionals with this form necessarily implies important dynamical properties of the network $\mathcal{G}$.

Take $(x(t) : t \geq 0)$ an orbit under $F_{\mathcal{G}}$ i.e. it satisfies equation (5.1). Assume the matrix of interactions A to be symmetric. Put:

$$E_{\mathcal{G}}(x(t-1)) = - < x(t), Ax(t-1) > + < b, x(t) + x(t-1) > \tag{5.4}$$

Let us evaluate the difference:

$$\Delta_{t-1} E_{\mathcal{G}} = E_{\mathcal{G}}(x(t-1)) - E_{\mathcal{G}}(x(t-2))$$

From symmetry on A we deduce:

$$\Delta_{t-1} E_{\mathcal{G}} = - < x(t) - x(t-2), Ax(t-1) - b >$$

or equivalently:

$$\Delta_{t-1} E_{\mathcal{G}} = - < f(Ax(t-1) - b) - f(Ax(t-3) - b), \; Ax(t-1) - b > \qquad (5.5)$$

Now we shall define the relevant class of functions for which it is direct to show that $E_{\mathcal{G}}(x(t-1))$ is decreasing.

Definition 5.1. (i) Let $\mathcal{D} \subset \mathbb{R}^n$ be convex. A function $f : \mathcal{D} \to \mathbb{R}^n$ is called *positive* iff:

$$< f(x) - f(y), x >\, \geq 0 \quad \text{for any } x, y \in \mathbb{R}^n \qquad (5.6)$$

(ii) A generalized network $\mathcal{G} = (I, A, b, C, f)$ is called *positive* if the function f is positive. ∎

From the above analysis we deduce:

Proposition 5.1. [G1] Let $\mathcal{G} = (I, A, b, C, f)$ be a positive network with a symmetric matrix of interactions A. Then the functional:

$$E_{\mathcal{G}}(x(t-1)) = - < x(t), Ax(t-1) > + < b, x(t) + x(t-1) >$$

is decreasing, and hence is a Lyapunov functional.

Proof. It follows from expressions (5.5), (5.6). ∎

To obtain dynamical consequences of Proposition 5.1, let us introduce the following stronger condition.

A positive function $f : \mathcal{D} \to \mathbb{R}^n$ is said to be *strictly positive* if it also satisfies the property:

$$\text{for } u, v \in \mathcal{D} : \quad < f(x) - f(y), x >\, = 0 \text{ iff } f(x) = f(y) \text{ or } x = 0 \qquad (5.7)$$

A generalized network $\mathcal{G} = (I, A, b, C, f)$ is called *strictly positive* if the function f is strictly positive.

Corollary 5.1. [G1] Let $\mathcal{G} = (I, A, b, C, f)$ be a strictly positive network with symmetric interactions A. Then any finite orbit $(x(t) : t \in \mathbb{Z}_T)$ has a period $T \leq 2$.

Proof. $E_{\mathcal{G}}$ being a Lyapunov functional it must be constant on the orbit so $\Delta_{t-1} E_{\mathcal{G}} = 0$ for any $t \in \mathbb{Z}_T$. Expressions (5.5), (5.7) imply $f(Ax(t-1) - b) = f(Ax(t-3) - b)$ or $Ax(t-1) - b = 0$ which is equivalent to $x(t) = x(t-2)$ or $Ax(t-1) - b = 0$. If for some $t \in \mathbb{Z}_T$ the equality $x(t) = x(t-2)$ holds, we conclude that the length of the orbit satisfies $T \leq 2$. If for any $t \in \mathbb{Z}_T$ the expression $Ax(t-1) - b = 0$ is satisfied, the orbit reduces to a fixed point. Then the corollary follows. ∎

Let us give some relevant examples of positive functions.

Example 5.1. Let $f_0 : \mathbb{R} \to \mathbb{R}$ be a real function. Then it is easy to show that f_0 is positive iff f_0 is of the form:

$$f_0(u) = \begin{cases} c_1 & \text{if } u < 0 \\ c_2 & \text{if } u = 0 \\ c_3 & \text{if } u > 0 \end{cases} \qquad \text{with } c_1 \leq c_2 \leq c_3$$

Recall that the threshold and the sign functions are positive real functions. It is direct to prove that any positive real function is strictly positive.

Obviously if $f : \mathbb{R}^{|I|} \to \mathbb{R}^{|I|}$ satisfies $f(x_i : i \in I) = (f_0(x_i) : i \in I)$, with f_0 a positive real function, this implies f is a strictly positive function. Then the Neural Networks are strictly positive networks because the function f is in this case either the multidimensional threshold or the sign function. ∎

Example 5.2. To any $x \in \mathbb{R}^{|I|}$ associate the index set:

$$C(x) = \{i \in I : x_i \geq x_j \text{ for any } j \in I\}$$

Let $e_i = (0, ..., 1, ..., 0)$ with 1 in the i-th coordinate. It can be easily shown that the function:

$$f_{Maj} : \mathbb{R}^{|I|} \to \mathbb{R}^{|I|}, \quad f_{Maj}(x) = \frac{1}{|C(x)|} \sum_{i \in C(x)} e_i$$

is positive.

Recall that f_{Maj} only charges the coordinates where the maximum is attained, so the network $\mathcal{G} = (I, A, b, C, f_{Maj})$ evolves as a majority network. The configuration space is the finite set $C = \{x \in \mathbb{R}^n : x_i = \frac{\ell_i}{n}, \ell_i \in \mathbb{N}, \sum_{i \in I} \ell_i = n\}$. ∎

An interesting example of positive functions whose range is a continuous set is the following:

Example 5.3. Let $p > 1$. On $\mathcal{D} = I\!R_+^n = \{x : x_i \geq 0 \text{ for any } 1 \leq i \leq n\}$ consider the p-norm:

$$\|x\|_p = \sum_{i=1}^n |x_i|^p \text{ for } x \in I\!R_+^n$$

Note $x^p = (x_i^p : i = 1, ..., n)$. Take $\mathcal{C} = \{x \in I\!R^n : \|x\|_p = 1 \text{ or } x = 0\}$. We shall prove that the following function $f : I\!R_+^n \to \mathcal{C}$ is strictly positive:

$$f(x) = \begin{cases} \frac{x^{p-1}}{\|x\|_p^{p-1}} & \text{if } x \neq 0 \\ 0 & \text{if } x = 0 \end{cases}$$

We have $< f(x) - f(y), x > = \|x\|_p - \frac{1}{\|y\|_p^{p-1}} < x, y^{p-1} >$. Apply Hölder inequality to x, y^{p-1} and coefficients $p, q = \frac{p}{p-1}$ to obtain:

$$< x, y^{p-1} > \leq \|x\|_p \|y^{p-1}\|_{\frac{p}{p-1}}$$

Since $\|y^{p-1}\|_{\frac{p}{p-1}} = \|y^p\|_p^{p-1}$ we conclude that f is positive.

Now assume $< f(x) - f(y), x > = 0$. From the above calculations we deduce $< x, y^{p-1} > = \|x\|_p \|y^{p-1}\|_{\frac{p}{p-1}}$; this means that the Hölder inequality becomes an equality. A necessary and sufficient condition that this occur is the existence of λ, λ' different from 0 satisfying $\lambda x^p = \lambda'(y^{p-1})^{\frac{p}{p-1}}$, hence $\lambda x^p = \lambda' y^p$. Since $x, y \in I\!R_+^n$ we deduce $f(x) = f(y)$. ∎

We shall describe the class of positive functions by using current notions of convex analysis theory. This will also give some light on the Lyapunov functional E_g.

Recall that a function $g : \mathcal{D} \to I\!R$ defined on a convex set $\mathcal{D} \subset I\!R^n$ is said to be:

convex if $g(\lambda x + (1 - \lambda)y) \leq \lambda g(x) + (1 - \lambda)g(y)$ for $\lambda \in [0, 1], x, y \in \mathcal{D}$ (5.8)

If $\mathcal{D}$ is also a *cone*, i.e. $\lambda x \in \mathcal{D}$ for any $\lambda \geq 0$ and $x \in \mathcal{D}$, we say it is:

positive homogeneous if $g(\lambda x) = \lambda g(x)$ for all $\lambda \geq 0, x \in \mathcal{D}$ (5.9)

Take a function $g : \mathcal{D} \to I\!R$. Then it is convex iff there exists a function $\gamma : \mathcal{D} \to I\!R^n$ satisfying:

$$g(x) \geq g(y) + < \gamma(y), x - y > \text{ for any } x, y \in \mathcal{D}$$

Any γ verifiying the above property is called a *subgradient* of g [Rck].

Now associate to any $f : D \to \mathbb{R}^n$ the following real functional:

$$\phi_f(x) = < f(x), x > \qquad \text{for } x \in D \tag{5.10}$$

We have the characterization:

Theorem 5.1. [GM2] Let $D \subset \mathbb{R}^n$ be a convex cone and $f : D \to \mathbb{R}^n$. Then the three following conditions are equivalent:

(i) f is a positive function.

(ii) $\phi_f(x) = < f(x), x >$ is a positive homogeneous convex function.

(iii) f is a subgradient of some positive homogeneous convex function.

If the above equivalent conditions are satisfied f is a subgradient of ϕ_f.

Proof. Assume that f is positive. Since condition $< f(x) - f(y), x > \geq 0$ is equivalent to $\phi_f(x) \geq \phi_f(y) + < f(y), x - y >$ we conclude that ϕ_f is convex and f is a subgradient of ϕ_f. From the inequalities $< f(\lambda x) - f(x), \lambda x > \geq 0$, $< f(x) - f(\lambda x), x > \geq 0$ we deduce $\phi_f(\lambda x) = \lambda \phi_f(x)$ for any $\lambda \geq 0$, $x \in \mathbb{R}^n$, so ϕ_f is positive homogeneous.

Now assume that f is a subgradient of some positive homogeneous convex function g. We shall prove that we necessarily have $g = \phi_f$, then the theorem will be shown.

Since g is convex positive homogeneous, we have

$$2g(x) = g(2x) \geq g(x) + < f(x), x >$$

On the other hand, g positive homogeneous implies $g(0) = 0$, therefore

$$0 = g(0) \geq g(x) + < f(x), -x >$$

The above two inequalities imply $g(x) = < f(x), x > = \phi_f(x)$. ∎

For any $f : \mathbb{R}^n \to \mathbb{R}^n$ we denote $\tilde{f}$ the function

$$\tilde{f}(x) = \frac{1}{2}(f(x) - f(-x)) \tag{5.11}$$

We call f *antisymmetric* iff $\tilde{f} = f$. This condition is equivalent to: $f(-x) = -f(x)$ for $x \in \mathbb{R}^n$. Obviously the function $\tilde{f}$ is antisymmetric.

From the expression:

$$2 < \tilde{f}(x) - \tilde{f}(y), x > = < f(x) - f(y), x > + < f(-x) - f(-y), -x >$$

we deduce that f positive implies $\tilde{f}$ positive. From the foregoing result we can deduce:

Corollary 5.2. [GM2] Let $f : \mathbb{R}^n \to \mathbb{R}^n$ be positive then $\phi_f(x) = <\tilde{f}(x), x>$ is a seminorm on $\mathbb{R}^n$. The functional ϕ_f is a norm iff the linear subspace generated by $\tilde{f}(\mathbb{R}^n)$ is $\mathbb{R}^n$.

Reciprocally, if g is a seminorm then some subgradient of its is an antisymmetric positive function.

Proof. Recall $\tilde{f}(0) = 0$ then $\phi_f(x) = <\tilde{f}(x) - \tilde{f}(0), x> \geq 0$. On the other hand the subadditive condition $\phi_f(x + y) \leq \phi_f(x) + \phi_f(y)$ follows from the convexity and the positive homogeneous properties of ϕ_f. Now $\tilde{f}$ antisymmetric implies $\phi_f(\lambda x) = |\lambda|\phi_f(x)$. Hence ϕ_f is a seminorm.

Now assume $\phi_f(x) = 0$. This implies $< \tilde{f}(y), x > = < \tilde{f}(x) - \tilde{f}(y), x > \geq 0$ and similarly $< \tilde{f}(-y), x > = < \tilde{f}(x) - \tilde{f}(y), x > \geq 0$. Then $< \tilde{f}(y), x > = 0$ for any $y \in \mathbb{R}^p$, so $x \perp \tilde{f}(\mathbb{R}^n)$. Reciprocally if $x \perp \tilde{f}(\mathbb{R}^n)$ we deduce $\phi_f(x) = < \tilde{f}(x), x > = 0$. Hence ϕ_f is a norm iff the linear subspace generated by $\tilde{f}(\mathbb{R}^n)$ is $\mathbb{R}^n$.

Now suppose g is a seminorm. Take f a subgradient of g. From the last theorem f is positive and $g(x) = < f(x), x >$. Consider the positive antisymmetric function $\tilde{f}$ defined in (5.11). It suffices to show that $\tilde{f}$ is also a subgradient of g.

From $g(x) = g(-x)$ we deduce $< f(x), x > = < f(-x), -x >$, so $g(x) = < \tilde{f}(x), x >$. Since $\tilde{f}$ is positive we conclude that it is a subgradient of g. ∎

In the above examples, let us associate the positive homogeneous convex function ϕ_f to the corresponding f:

In example 5.1, the convex potential ϕ_f is

$$\phi_f(u) = \begin{cases} a_1 u & \text{if } u < 0 \\ 0 & \text{if } u = 0 \\ a_2 u & \text{if } u = 0 \end{cases}$$

Recall that ϕ_f is a seminorm in the antisymmetric case, i.e. iff $a_3 = -a_1 \geq 0$, and it is a norm if also $a_3 > 0$.

In particular for the antisymmetric function $f = \text{sign}$ we have $\phi_{sign}(u) = |u|$ the current norm in $\mathbb{R}$. For the threshold function $f = \mathbb{1}$ we get

$$\phi_{\mathbb{1}}(u) = u^+ = \begin{cases} u & \text{if } u \geq 0 \\ 0 & \text{if not} \end{cases}$$

In example 5.2 we get:

$$\phi_{f_{Maj}}(x) = < f_{Maj}(x), x > = \frac{1}{|C(x)|} \sum_{i \in D(x)} < e_i, x > = \max(x_i : i \in I)$$

In example 5.3 the function f is antisymmetric and

$$\phi_f(x) = \begin{cases} < \frac{x^{p-1}}{\|x\|_p^{p-1}}, x > & \text{if } x \neq 0 \\ 0 & \text{if } x = 0 \end{cases}$$

Hence $\phi_f(x) = \|x\|_p$ is the p-norm.

Now recall that the Lyapunov functional E_g defined in (5.4) can also be written as:

$$E_g(x(t-1)) = - < f(Ax(t-1) - b), Ax(t-1) - b > + < b, x(t-1) >$$

Or, with the above notation:

$$E_g(x(t-1)) = -\phi_f(Ax(t-1) - b) + < b, x(t-1) >$$

Since f is positive the expression $\phi_f(Ax(t-1) - b)$ is a positive homogeneous convex function which is also a seminorm if f is antisymmetric.

Assume the vector $b = 0$. In this case the Lyapunov functional takes the form: $E_g(x(t-1)) = -\phi_f(Ax(t-1))$. We get the following relationship between optimization problems and automata network evolution:

Corollary 5.3. [GM2] Let $g : D \to \mathbb{R}$ be a convex positive homogenous function (for instance g a seminorm) and $f : \mathbb{R}^n \to C$ be a subgradient of g. Then any solution of the maximization problem:

$$g(Ay_0) = \max_{y \in C} g(Ay), \quad y_0 \in C \qquad (5.12)$$

belongs to a fixed point or a period-2 orbit of the positive automaton $\mathcal{G} = (I, A, b = 0, C, f)$.

Proof. $g(Ax(t-1)) = \phi_f(Ax(t-1))$ is strictly increasing in the transient. So any solution y_0 necessarily belongs to a cycle. ∎

In this context the synchronous dynamics is a "Hill-Climbing" strategy to solve the optimization problems 5.12.

For special cases we shall be able to get a finer version of the above result, as in section 5.6.

5.3. Multithreshold Networks

A Multithreshold Automata network behaves like an ordinary Neural Network but instead of being constrained to take only the state values 0 or 1 it can take a value varying in a finite real set.

To be more precise let $Q = \{q^{(0)}, ..., q^{(p-1)}\} \subset \mathbb{R}$ be a finite set of real values, indexed in a monotonically increasing way: $q^{(k)} < q^{(k+1)}$ for $k = 0, ..., p-2$. Consider a multithreshold vector $b = (b^{(k)} : k = 1, ..., p-1)$ of increasing values: $b^{(k)} < b^{(k+1)}$ for $k = 1, ..., p-2$. Also define $b^{(0)} = -\infty$, $b^{(p)} = \infty$. The multithreshold function $M_b : \mathbb{R} \to Q$ is given by (see Figure 5.1):

$$M_b(u) = q^{(k)} \quad \text{iff} \quad b^{(k)} \le u < b^{(k+1)} \tag{5.13}$$

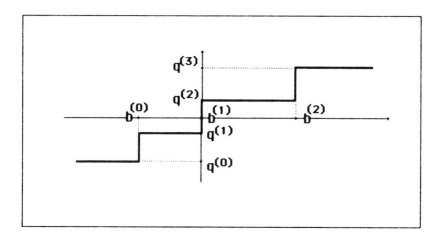

Figure 5.1. M_b for $p = 4$.

Recall that if $Q = \{0, 1\}$ then $b \in \mathbb{R}$ and $M_b(u) = \mathbb{1}(u - b)$ for any $u \in \mathbb{R}$.

For I a set and $B = (b_i : i \in I)$ a matrix of multithreshold vectors, i.e. $b_i = (b_i^{(k)} : k = 1, ..., p-1)$, we define the generalized multithreshold function $\bar{M}_B : \mathbb{R} \to Q^{|I|}$ by $\bar{M}_B(u_i : i \in I) = (M_{b_i}(u_i) : i \in I)$.

Now take $I = \{1, ..., n\}$ a finite set of cells, $Q = \{q^{(0)}, ..., q^{(p-1)}\}$ an ordered subset of real values, $A = (a_{ij} : i, j \in I)$ a matrix of connections and

$B = (b_i : i \in I)$ a matrix of multithresholds. The tuple $M = (I, Q, A, B)$ is called
a multithreshold network and its dynamics is given by the transformation:

$$F_M(x) = \bar{M}_B(Ax) \quad \text{for any } x \in Q^n \tag{5.14}$$

hence any trajectory $(x(t) : t \geq 0)$ satisfies:

$$x(t+1) = \bar{M}_B(Ax(t)) \text{ i.e.:}$$
$$x_i(t+1) = q^{(k)} \quad \text{iff} \quad b_i^{(k)} \leq \sum_{j \in I} a_{ij} x_j(t) < b_i^{(k+1)} \tag{5.15}$$

When the matrix A is symmetric we call $M = (I, Q, A, B)$ a symmetrical
multithreshold network.

If $Q = \{0, 1\}$ we get the usual $b_i \in I\!R$ for any $i \in I$ so $F_M(x) = \bar{M}_B(Ax) =$
$\bar{\mathbb{1}}(Ax - B)$. Hence Multithreshold Networks are in principle more general than
finite Neural Networks. But we shall prove that any multithreshold network can
be simulated with some coding by a finite Neural Network, hence all the dynamical
properties of these two kind of networks are the same. Also we shall show that
the coding respects symmetry of the connection matrix, so the symmetric Multithreshold Networks will possess exactly the same properties as the symmetric
Neural Networks.

Theorem 5.2. [GM1] Let $M = (I, Q, A, B)$ be a Multithreshold Network. Then
there exists a finite Neural Network $N = (\bar{I}, \bar{A}, \bar{b})$ and a one-to-one encoding
function:

$$\varphi : Q^{|I|} \to \{0, 1\}^{|\bar{I}|}$$
$$\text{such that} \quad F_N \circ \varphi = \varphi \circ F_M \tag{5.16}$$

where $F_M(x) = \bar{M}_B(Ax)$ and $F_N(\bar{x}) = \bar{\mathbb{1}}(\bar{A}\bar{x} - \bar{b})$ are the evolution functions of
M and N respectively.

Furthermore the above construction preserves symmetry:

$$A \text{ symmetric} \implies \bar{A} \text{ symmetric}$$

Then if M is a symmetric Multithreshold Network the Neural Network N is also
symmetric.

Proof. First let us show that any multithreshold network $M = (I, Q, A, B)$ can be
put into equivalence with a multithreshold network $M' = (I, Q', A, B')$ satisfying
$|Q'| = |Q|$, $Q' = \{q'^{(0)}, ..., q'^{(p-1)}\} \subset I\!R_+$ and $q'^{(0)} = 0$.

For any $c \in \mathbb{R}$ take $Q' = Q + c = \{q^{(0)} + c, ..., q^{(p-1)} + c\}$,
$B' = \{b_i'^{(k)} = b_i^{(k)} + \sum_{j \in I} a_{ij}c : i \in I,\ k = 1, ..., p-1\}$. Now consider the automaton
$M' = (I, Q', A, B')$. For proving that M and M' are equivalent define the functions:

$$\psi_0 : Q \to Q' \quad \text{with } \psi_0(q^{(k)}) = q^{(k)} + c \quad \text{and}$$
$$\psi : Q^n \to Q'^n \quad \text{such that } \psi(q^{(k_i)} : i \in I) = (\psi_0(q^{(k_i)}) : i \in I)$$

It is easy to show that ψ is a one-to-one onto function satisfying:
$\psi \circ F_M = F_{M'} \circ \psi$. Hence the multithreshold networks M and M' are equivalent
(this equivalence relation is the canonical one among dynamical systems), so the
dynamical properties of M and M' are the same.

By taking $c = -q(0)$ we get $q'^{(0)} = 0$ and $q'^{(k)} > 0$ for any $k = 1, , ..., p-1$
so we have have shown our assertion. Recall that the interaction matrix A is the
same for M and M'; hence the above construction preserves symmetry.

Therefore we can restrict ourselves to construct $N = (I, \bar{A}, \bar{b})$ and the encoding
function φ of the theorem for the multithreshold networks $M = (I, Q, A, B)$ which
satisfy $Q = \{q^{(0)}, ..., q^{(p-1)}\} \subset \mathbb{R}_+$ and $q^{(0)} = 0$. For general multithreshold
networks M, we must compose with the equivalence constructed above to obtain
N and φ.

Denote $I = \{1, ..., n\}$, take $\bar{I} = \{1, ..., n(p-1)\}$. Define:
$\varphi_0 : Q \to \{0,1\}^{p-1}$ with $\varphi_0(q^{(k)}) = (y_\ell^{(k)} : \ell = 1, ..., p-1)$ such that:

$$y_\ell^{(k)} = \begin{cases} 1 & \text{if } \ell < k \\ 0 & \text{if } \ell \geq k \end{cases}$$

For $\tilde{q} \in Q$ we denote $K(\tilde{q}) = k$ if $\tilde{q} = q_k$. Then for $y = (y^{(\ell)} : \ell = 1, ..., p-1) = \varphi_0(\tilde{q})$ we have $K(\tilde{q}) = \sum_{\ell=1}^{p-1} y^{(\ell)}$. We write $K(y) = K(\tilde{q})$.

We take the encoding function $\varphi : Q^n \to \{0,1\}^{n(p-1)}$ so as to have
$\varphi(x_i : i \in I) = (\varphi_0(x_i) : i \in I)$.

Define the symmetric matrix $C = (c_{\ell r} : 1 \leq \ell, r \leq p-1)$, where:

$$c_{\ell r} = (q^{(\ell)} - q^{(\ell-1)})(q^{(r)} - q^{(r-1)}) \quad \text{for } \ell,\ r = 1, ..., p-1$$

Let $y = (y^{(r)} : r = 1, ..., p-1) = \varphi_0(q^{(k)})$. The ℓ-th term of the vector Cy
is: $(Cy)_\ell = \sum_{r=1}^{p-1} c_{\ell r} y^{(r)} = (q^{(\ell)} - q^{(\ell-1)}) \sum_{r=1}^{p-1} (q^{(r)} - q^{r-1}) y^{(r)} = (q^{(\ell)} - q^{(\ell-1)}) q^{(k)}$.
Hence:

$$C\varphi_0(q^{(k)}) = q^{(k)}(q^{(\ell)} - q^{(\ell-1)}) : \ell = 1, ..., p-1) \quad \text{for } k = 0, ..., p-1 \quad (5.17)$$

Now define the $n(p-1) \times n(p-1)$ matrix:

$$\bar{A} = \begin{matrix} \bar{A}_{11} & \cdot & \bar{A}_{1j} & \cdot & \bar{A}_{1n} \\ \cdot & \cdot & \cdot & \cdot & \cdot \\ \bar{A}_{i1} & \cdot & \bar{A}_{ij} & \cdot & \bar{A}_{in} \\ \cdot & \cdot & \cdot & \cdot & \cdot \\ \bar{A}_{n1} & \cdot & \bar{A}_{nj} & \cdot & \bar{A}_{nn} \end{matrix} \quad \text{where } \bar{A}_{ij} = a_{ij}C \text{ is an } (p-1) \times (p-1) \text{ matrix.}$$

C being symmetric we find that: A symmetric implies $\bar{A}$ symmetric.

The vector $\bar{b} = (\bar{b}_i^{(k)} : k = 1, ..., p-1; i \in I)$ satisfies $\bar{b}_i^{(k)} = b_i^{(k)}(q^{(k)} - q^{(k-1)})$. For $x = (x_i : i \in I) \in Q^n$ we have $\varphi(x) = (\varphi_0(x_i) : i \in I)$. Let $z = (z_i^{(\ell)} : \ell = 1, ..., p-1; i \in I) = \bar{A}\bar{\varphi}(x)$. From (5.17) we obtain:

$$z_i^{(\ell)} = \sum_{j \in I} a_{ij}(C\varphi_0(x_j))^{(\ell)} \sum_{j \in I} a_{ij}x_j(q^{(\ell)} - q^{(\ell-1)})$$

Then $z_i^{(\ell)} \geq b_i^{(\ell)}(q^{\ell} - q^{(\ell-1)})$ iff $\sum_{j \in I} a_{ij}x_j \geq b_i^{(\ell)}$. From the definition of $\bar{b}$ we get: $\mathbb{1}(\bar{A}\varphi(x) - \bar{b}) = \varphi(Ax - b)$. We have shown the theorem. ∎

The above construction may be illustrated with the following example. Consider the multithreshold network given by $I = \{1, 2\}$, $Q = \{0, 1, 2, 3\}$, $A = \binom{0\,1}{1\,0}$, $B = \binom{1\,4\,5}{2\,3\,5}$. Note $F_N(x_1, x_2) = (F_1(x_1, x_2), F_2(x_1, x_2))$. We have:

$$F_1(x_1, x_2) = \begin{cases} 0 & \text{if} & x_2 < 1 \\ 1 & \text{if} & 1 \leq x_2 < 4 \\ 2 & \text{if} & 4 \leq x_2 < 5 \\ 3 & \text{if} & 5 \leq x_2 \end{cases} \quad \text{and} \quad F_2(x_1, x_2) = \begin{cases} 0 & \text{if} & x_1 < 2 \\ 1 & \text{if} & 2 \leq x_1 < 3 \\ 2 & \text{if} & 3 \leq x_1 < 5 \\ 3 & \text{if} & 5 \leq x_1 \end{cases}$$

The connection graph is exhibited in Figure 5.2.i.

By coding $0 \to (000)$, $1 \to (100)$, $2 \to (110)$, $3 \to (111)$ we obtain the graph of Figure 5.2.ii with binary threshold rules. Note $(F_N(x))_{(i_1, i_2)} = F_{(i_1, i_2)}(x)$ for $1 \leq i_1 \leq 2$, $1 \leq i_2 \leq 3$:

$$F_{11}(x_{21}, x_{22}, x_{23}) = \mathbb{1}(\sum_{j=1}^3 x_{2j} - 1), F_{12}(x_{21}, x_{22}, x_{23}) = \mathbb{1}(\sum_{j=1}^3 x_{2j} - 4)$$

$$F_{13}(x_{21}, x_{22}, x_{23}) = \mathbb{1}(\sum_{j=1}^3 x_{2j} - 5), F_{21}(x_{11}, x_{12}, x_{13}) = \mathbb{1}(\sum_{j=1}^3 x_{1j} - 2)$$

$$F_{22}(x_{11}, x_{12}, x_{13}) = \mathbb{1}(\sum_{j=1}^3 x_{1j} - 3), F_{23}(x_{11}, x_{12}, x_{13}) = \mathbb{1}(\sum_{j=1}^3 x_{1j} - 5)$$

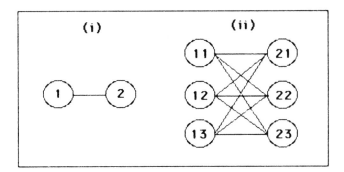

Figure 5.2. (i) Multithreshold graph. (ii) Binary threshold graph.

Corollary 5.4. The steady states of symmetric Multithreshold Networks are periodic points of period $T \leq 2$.

Proof. Assume that the point $x \in Q^n$ belongs to a periodic orbit of period T. Let $\mathcal{N}$ and φ be the Neural Network and the encoding function constructed in the last theorem. Hence $F_{\mathcal{N}}^T(\varphi(x)) = \varphi(F_{\mathcal{M}}^T(x)) = \varphi(x)$, so $\varphi(x)$ is a periodic point of period T. Since $\mathcal{N}$ is also symmetric we conclude by Theorem 2.1. that $T \leq 2$. ∎

Corollary 5.4 has also been proved in the framework of algebraic invariants by using the combinatorial notion of monotonic chains [GO1].

Also by using the encoding function φ we can associate to any multithreshold network $\mathcal{M} = (I, Q, A, B)$ a Lyapunov functional. To this purpose take $\{x(t)\}_{t \geq 0}$ an orbit of $\mathcal{M}$, i.e. $x(t+1) = F_{\mathcal{M}}(x(t))$. Let $\varphi : Q^n \to \{0,1\}^{n(p-1)}$ be the encoding between $\mathcal{M}$ and $\mathcal{N} = (\bar{I}, \bar{A}, \bar{b})$, so $\varphi(x(t+1)) = F_{\mathcal{N}}(\varphi(x(t)))$ and we conclude $\{\bar{x}(t) = \varphi(x(t))\}_{t \geq 0}$ is an orbit of $\mathcal{N}$.

When A is symmetric we get $\bar{A}$ symmetric, in this case Proposition 3.1. asserts $E(\bar{x}(t)) = - < \bar{x}(t), \bar{A}x(t-1) > + < \bar{b}, \bar{x}(t) + \bar{x}(t-1) >$ is a decreasing functional for the orbit $\{\bar{x}(t)\}_{t \geq 0}$. Then:

$$E_{\mathcal{M}}(x(t)) = - < \varphi(x(t)), \bar{A}\varphi(x(t-1)) > + < \bar{b}, \varphi(x(t)) + \varphi(x(t-1)) > \quad (5.18)$$

is a Lyapunov functional for the multithreshold automata $\mathcal{M}$ where $\varphi, \bar{A}, \bar{b}$ are, respectively, the encoding function, the matrix of connections and the threshold vectors constructed in the proof of Theorem 5.2. Recall that the above inner product $<,>$ is the standard one in $\mathbb{R}^{n(p-1)}$

Now we shall express the first part of the Lyapunov functional $E_{\mathcal{M}}$ in terms of the matrix A. We assume $q^{(0)} = 0$.

Denote:

$$\varphi(x(t)) = (y_i^{(k)} : i \in I, \ k = 1, ..., p-1),$$
$$\bar{A}\varphi(x(t-1)) = (z_i^{(k)} : i \in I, \ k = 1, ..., p-1).$$

So:

$$< \varphi(x(t)), \bar{A}\varphi(x(t-1)) >= \sum_{i \in I} \sum_{k=1}^{p-1} y_i^{(k)} z_i^{(k)}$$

The last part in the proof of theorem 5.2. implies

$$z_i^{(k)} = (q^{(k)} - q^{(k-1)}) \sum_{j \in I} a_{ij} x_j(t-1)$$

By definition of φ we also have:

$$\sum_{k=1}^{p-1} (q^{(k)} - q^{(k-1)}) y_i^{(k)} = x_i(t)$$

Use these equalities to obtain:

$$< \varphi(x(t)), \bar{A}\varphi(x(t-1)) >= \sum_{i \in I} a_{ij} x_j(t-1) x_i(t) =< x(t), Ax(t-1) >$$

where the latter inner product is the usual one in $I\!\!R^n$.

To examine the second part of the Lyapunov functional recall that for any $y = (y^{(\ell)} : \ell = 1, ..., p-1) = \varphi_0(\tilde{q})$ we denoted $K(\tilde{q}) = K(y) = \sum_{\ell=1}^{p-1} y^{(\ell)}$. This integer satisfied $\varphi_0(q_{K(y)}) = y$.

If $\bar{y} = (y_1, ..., y_n) \in \varphi(Q)$ we have: $< \bar{b}, \bar{y} >= \sum_{i \in I} \sum_{k=1}^{p-1} \bar{b}_i^{(k)} y_i^{(k)} =$ $\sum_{i \in I} \sum_{k=1}^{K(y)} \bar{b}_i^{(k)}$. Hence:

Corollary 5.5. For Multithreshold Networks $\mathcal{M} = (I, Q, A, B)$ satisfying $q^{(0)} = 0$ the function:

$$E_{\mathcal{M}}(x(t)) = - < x(t), Ax(t-1) > + \sum_{i \in I} \left(\sum_{k=1}^{K(x_i(t))} \bar{b}_i^{(k)} + \sum_{k=1}^{K(x_i(t-1))} \bar{b}_i^{(k)} \right) \quad (5.19)$$

is a Lyapunov functional for their dynamics. ∎

5.4. Approximation of Continuous Networks by Multithreshold Networks

Let $f_0 : \mathbb{R} \to \mathbb{R}$ be a real functional satisfying the hypothesis:

$$f_0(0) = 0, \; f_0(c) = c, \; f_0 \text{ is continuous strictly increasing in } [0, c]$$
$$f_0(u) = 0, \text{ if } u < 0, \; f_0(u) = c \text{ if } u > c \tag{5.20}$$

For any $p \geq 2$ divide $[0, c]$ into pieces of length $\frac{c}{p-1}$ and denote $Q_{(p)}$ the set of their extremities:

$$Q_{(p)} = \{u_{(p)}^{(k)} = \frac{k}{p-1} c : k = 0, ..., p-1\} \tag{5.21}$$

Also denote by $\varphi_{(p)}$ the following function which sends $\mathbb{R}$ into $Q_{(p)}$:

$$\varphi_{(p)}(u) = \begin{cases} u_{(p)}^{(0)} = 0 & \text{if } u < u_{(p)}^{(1)} = \frac{c}{p-1} \\ u_{(p)}^{(k)} & \text{if } u_{(p)}^{(k)} \leq u < u_{(p)}^{(k+1)} \\ u_{(p)}^{(p-1)} = c & \text{if } u \geq u_{(p)}^{(p+1)} = c \end{cases} \tag{5.22}$$

We have:

$$\varphi_{(p)}(u) \xrightarrow[u \to \infty]{} u \text{ uniformly in } [0, c] \tag{5.23}$$

Now define the points:

$$\gamma_{(p)}^{(k)} = f_0^{-1}(u_{(p)}^{(k)}) \quad \text{for } k = 1, ..., p-1$$
$$\gamma_{(p)}^{(0)} = -\infty, \; \gamma_{(p)}^{(p)} = \infty \tag{5.24}$$

The hypothesis satisfied by f_0 imply:

$$\varphi_{(p)}(f_0(u)) = u_{(p)}^{(k)} \quad \text{iff } \gamma_{(p)}^{(k)} \leq x < \gamma_{(p)}^{(p+1)} \tag{5.25}$$

Let $I = \{1, ..., n\}$ be a set of cells, A a connection matrix, $b = (b_i \in I)$ a vector. Take $Q = [0, c]$ and consider the generalized continuous network $\mathcal{G} = (I, A, b, Q, f)$ where the function $f : \mathbb{R}^n \to \mathbb{R}^n$ is defined by $f(x_i : i \in I) = (f_0(x_i) : i \in I)$; thus the evolution of the network is $F_{\mathcal{G}}(x) = f(Ax - b)$.

We shall associate to the continuous network $\mathcal{G}$ a sequence of Multithreshold Networks $M_{(p)} = (I, A, B_{(p)}, Q_{(p)})$ where $Q_{(p)}$ is given by (5.21) and the threshold matrix $B_{(p)} = (b_i^{(k)} : i \in I, \; k = 1, ..., p-1)$ satisfies:

$$b_i^{(k)} = \gamma_{(p)}^{(k)} + b_i \tag{5.26}$$

As usual we put $b_i^{(0)} = -\infty$, $b_i^{(p)} = \infty$.

The transformation $F_{M_{(p)}}$ of this network is:

$$(F_{M_{(p)}}(y))_i = (\bar{M}_{B_{(p)}}(Ay))_i = u_{(p)}^{(k)} \quad \text{iff } \gamma_{(p)}^{(k)} + b_i \le \sum_{j \in I} a_{ij} y_j < \gamma_{(p)}^{(k+1)} + b_i$$

From (5.25) we deduce:

$$(F_{M_{(p)}}(y))_i = u_{(p)}^{(k)} \quad \text{iff } \varphi_{(p)}(\sum_{j \in I} a_{ij} y_j - b_i) = u_{(p)}^{(k)} \tag{5.27}$$

Take $u \in Q = [0, c]$; then the point $\varphi_{(p)}(u)$ lies on $Q_{(p)}$. From (5.23) and (5.27) it follows:

$$F_{M_{(p)}}(\varphi_{(p)}(u)) \xrightarrow[p \to \infty]{} F_{\mathcal{G}}(u) \tag{5.28}$$

Then we can deduce that the trajectories of the Multithreshold Network $M_{(p)}$ approach the trajectories of the generalized continuous network $\mathcal{G}$.

More precisely let $x(0) \in Q^n$ and $(x(t) : t \ge 0)$ be its trajectory under $F_{\mathcal{G}}$, i.e. $x(t+1) = F_{\mathcal{G}}(x(t))$. Take $y_{(p)}(0) = (\varphi_p(x_i(0)) : i \in I)$ and consider its trajectory $(y_{(p)}(t) : t \ge 0)$, i.e. $y_{(p)}(t+1) = F_{M_{(p)}}(y_{(p)}(t))$. From (5.28) we obtain:

$$y_{(p)}(t) \xrightarrow[p \to \infty]{} x(t) \text{ for any } t, \text{ uniformly with respect to } x(0) \in Q \tag{5.29}$$

Assume the connection matrix A to be symmetric. Denote $K(i, p, t) = K((y_{(p)}(t))_i)$ (where $K(\cdot)$ was defined in the proof of Theorem 5.2). Consider the Lyapunov functional

$$E_{M_{(p)}}(y_{(p)}(t-1)) =$$

$$- < y_{(p)}(t), Ay_{(p)}(t-1) > + \sum_{i \in I} \left(\sum_{k=1}^{K(i,p,t)} \bar{b}_i^{(k)} + \sum_{k=1}^{K(i,p,t-1)} \bar{b}_i^{(k)} \right)$$
$$\tag{5.30}$$

We have $E_{M_{(p)}}(t) - E_{M_{(p)}}(t-1) \le 0$, so if the limits $E_\infty(t) = \lim_{p \to \infty} E_{M_{(p)}}(t)$ exist for any $t \ge 1$ we should deduce $E_\infty(t) - E_\infty(t-1) \le 0$ for any $t \ge 1$.

From (5.29) we have: $< y_{(p)}(t), Ay_{(p)}(t-1) > \xrightarrow[p \to \infty]{} < x(t), Ax(t-1) >$.

Let us examine the second part of $E_{M_{(p)}}$.

Assume $(y_{(p)})_i = u_{(p)}^{(k)} = \frac{k}{p-1} c$. From the definition of $K((y_{(p)})_i)$ we obtain $K((y_{(p)})_i) = k = (\frac{p-1}{c})(y_{(p)})_i$. Besides, $\bar{b}_i^{(k)} = b_i^{(k)}(q^{(k)} - q^{(k-1)})$. We have $q^{(k)} = \frac{kc}{p-1}$ so $q^{(k)} - q^{(k-1)} = \frac{c}{p-1}$. From (5.26): $\bar{b}_i^{(k)} = (\gamma_{(p)}^{(k)} + b_i)\frac{c}{p-1}$.

Hence:

$$\sum_{k=1}^{K(i,p,t)} \bar{b}_i^{(k)} = \sum_{k=1}^{K(i,p,t)} \gamma_{(p)}^{(k)} \frac{c}{p-1} + b_i \frac{c}{p-1} K(i,p,t)$$

$$= \sum_{k=1}^{(\frac{p-1}{c})(y_{(p)})_i} f_0^{-1}(u_{(p)}^{(k)}) \frac{c}{p-1} + b_i (y_{(p)})_i.$$

From (5.23), (5.24), (5.29) we obtain:

$$\sum_{k=1}^{K(i,p,t)} \bar{b}_i^{(k)} \xrightarrow[p\to\infty]{} \int_0^{x_i(t)} f_0^{-1}(u)\,du + b_i x_i(t)$$

So if we define:

$$E_g\left(x(t-1)\right) = - <x(t), Ax(t-1)> + <b, x(t) + x(t-1)> +$$

$$\sum_{i\in I}\left(\int_0^{x_i(t)} f_0^{-1}(u)\,du + \int_0^{x_i(t-1)} f_0^{-1}(u)\,du\right) \tag{5.31}$$

we deduce:

$$E_{M_{(p)}}\left(y_{(p)}(t-1)\right) \xrightarrow[p\to\infty]{} E_g\left(x(t-1)\right) \tag{5.32}$$

Based on the formula $\int_0^{f_0(v)} f_0^{-1}(u)\,du = vf_0(v) - \int_0^{v} f_0(u)\,du$ and the equality $x_i(t) = f_0\left(\sum_{j\in I} a_{ij}x_j(t-1) - b_i\right)$ we get the relationship:

$$\sum_{i\in I}\left(\int_0^{x_i(t)} f_0^{-1}(u)\,du + \int_0^{x_i(t-1)} f_0^{-1}(u)\,du\right) = <x(t), Ax(t-1)>$$

$$+ <x(t-1), Ax(t-2)> - <b_1 x(t) + x(t-1)>$$

$$- \int_0^{\sum_{j\in I} a_{ij}x_j(t-1)-b_i} f_0(u)\,du - \int_0^{\sum_{j\in I} a_{ij}x_j(t-2)-b_i} f_0(u)\,du$$

Then $E_{\mathcal{G}}\left(x(t-1)\right)$ can be written as:

$$E_{\mathcal{G}}\left(x(t-1)\right) = <x(t-1), Ax(t-2)> - $$

$$\sum_{i \in I} \left(\int_{0}^{\sum_{j \in I} a_{ij}x_j(t-1)-b_i} f(u)du + \int_{0}^{\sum_{j \in I} a_{ij}x_j(t-2)-b_i} f(u)du \right) \qquad (5.33)$$

As a conclusion to this discussion (which follows the arguments developed in [M3]), we have:

Proposition 5.2. [GM3] Assume $f_0(0) = 0$, $f_0(c) = c$, $f_0(u) = 0$ if $u < 0$, $f_0(u) = c$ if $u > c$ and f continuous strictly increasing in $[0, c]$. Consider the continuous network $\mathcal{G} = (I, A, b, Q = [0, c], f)$ with dynamics given by $F_{\mathcal{G}}(x) = f(Ax - b)$, and $(f(x) = f_0(x_i) : i \in I)$. If the matrix of connections A is symmetric the functional $E_{\mathcal{G}}(t)$ given by (5.31) (or (5.33)) is decreasing with t. ∎

5.5. Cyclically Monotone Networks

Here we consider Cyclically Monotone networks which contain the positive networks and the continuous newtoks introduced in the previous section. By using convex analysis concepts we hall define Lyapunov functionals for this class of networks thus allowing simple characterizations of the steady states.

Let $I = \{1, ..., n\}$ be the finite set of cells. The configuration set is $C \subset \mathbb{R}^n$. Let $A = (a_{ij} : i, j \in I)$ be the matrix of connections and $b = (b_i : i \in I)$ a vector. Assume $D \subset \mathbb{R}^n$ is a convex subset preserved by A, i.e. $A(D) \subset D$ and $f : AD \to C$ sends D on the configuration space C. This generalized network is denoted by $\mathcal{G} = (I, A, b, C, f)$. Its evolution is given by the dynamical equation:

$$x(t+1) = F_{\mathcal{G}}\left(x(t)\right), \text{ with } F_{\mathcal{G}} : C \to C \text{ defined by } F_{\mathcal{G}} = f(Ax - b) \qquad (5.34)$$

Definition 5.2. (i) Let D be convex. A function $f : D \to \mathbb{R}^n$ is *cyclically monotone* if for any cycle of vectors of length $m \geq 2$: $(y(s) \in D : s \in \mathbb{Z}_m)$, the inequality

$$\sum_{s=0}^{m-1} < f(y(s+1)) - f(y(s)), y(s+1) > \geq 0 \qquad (5.35)$$

holds.

(ii) A generalized network $\mathcal{G} = (I, A, b, C, f)$ is called *Cyclically Monotone* if $f : \mathcal{D} \to C$ is a cyclically monotone function.

From definition 5.1 it follows that positive functions are cyclically monotone, so many results we found for positive networks will be particular cases of those we shall establish in this section.

An equivalent characterization is the following. A function $f : \mathcal{D} \to \mathbb{R}^n$ is cyclically monotone iff it is the subgradient of some convex function $g : \mathcal{D} \to \mathbb{R}$ (see [Rck]). This means:

$$g(x) \geq g(y) + < f(y), x - y > \qquad \text{for every } x, y \in \mathcal{D} \qquad (5.36)$$

Any such convex g is called a (convex) potential of f.

For any convex function $g : \mathcal{D} \to \mathbb{R}$ defined on the closed convex set $\mathcal{D}$ we can define the *polar* function g^* associated to it by the formula:

$$g^*(y) = \sup\{< x, y > -g(x) : x \in \mathcal{D}\}, \quad y \in \mathcal{D} \qquad (5.37)$$

When f is a subgradient of g we find:

$$g^*(f(x)) + g(x) = < f(x), x > \qquad \text{for } x \in \mathcal{D} \qquad (5.38)$$

Now we have all the concepts to introduce Lyapunov functionals for this kind of networks.

Theorem 5.3. [GM3] Let $\mathcal{G} = (I, A, b, C, f)$ be a Cyclically Monotone network with symmetric matrix of connections A.

For any trajectory $(x(t)) : t \geq 0)$ the functional:

$$E_{\mathcal{G}}(x(t)) = g^*(x(t)) - g(Ax(t) - b) + < b, x(t) >$$

is decreasing with t, so it is a Lyapunov functional.

Proof. We shall study the difference $\Delta_t E_{\mathcal{G}} = E_{\mathcal{G}} - E_{\mathcal{G}}(t - 1)$. From (5.38) and symmetry we get:

$$g^*(x(t)) = g^*(f(Ax(t - 1) - b))$$
$$= < f(Ax(t - 1) - b), Ax(t - 1) - b > -g(Ax(t - 1) - b)$$
$$= < x(t), Ax(t - 1) > + < b, x(t) > -g(Ax(t - 1) - b)$$

Hence:

$$E_{\mathcal{G}}(x(t)) = < x(t - 1), Ax(t) > -g(Ax(t - 1) - b) - g(Ax(t) - b) \qquad (5.39)$$

So:

$$\Delta_t E_g = < f(Ax(t-2) - b), Ax(t) - b - (Ax(t-2) - b) > \\ + g(Ax(t-2) - b) - g(Ax(t) - b) \tag{5.40}$$

Denoting $u(t) = Ax(t) - b$ we can write:

$$\Delta_t E_g = g(u(t-2)) - g(u(t)) + < f(u(t-2)), u(t) - u(t-2) >$$

This expression is ≤ 0 because f is a subgradient of g. ∎

In order to deduce dynamical properties from the form of the functional E_g we impose a little stronger condition. Let f be a cyclically monotone function. From (5.35) it follows that $< f(x) - f(y), x - y >\geq 0$ for any $x, y \in \mathcal{D}$. We call f *strictly cyclically monotone* function if it also satisfies

$$< f(x) - f(y), x - y >= 0 \quad \text{iff } f(x) = f(y) \tag{5.41}$$

From the equality

$$< f(x) - f(y), x - y >=< f(x) - f(y), x > + < f(y) - f(x), y >$$

we deduce that a strictly positive function is strictly cyclically monotone.

We can give a characterization of strictly cyclically monotone functions in terms of their potentials.

Lemma 5.1. Let f be a cyclically monotone function and g a potential of f. Then f is strictly cyclically monotone iff:

$$g(y) = g(x) + < f(x), y - x > \quad \text{implies } f(x) = f(y) \tag{5.42}$$

Proof. From the inequalities $g(x) \geq g(y) + < f(y), x - y >$ and $g(y) \geq g(x) + < f(x), y - x >$ we deduce that $< f(x) - f(y), x - y >= 0$ iff $g(y) = g(x) + < f(x), y - x >, g(x) = g(y) + < f(y), x - y >$. Hence (5.42) implies f is strictly cyclically monotone.

Now suppose that property (5.41) holds. Pick x, y such that $g(y) = g(x) + < f(x), y - x >$. If $x \neq y$ take $z = \lambda y + (1 - \lambda)x$. We easily deduce $g(y) \geq g(x) + < f(x), z - x > + < f(z), y - z >$ so $< f(x), y - x >\geq< f(x), z - x > + < f(x), y - z >$. Then $< f(x) - f(z), z - y >\leq 0$. Since $z - y = (1 - \lambda)(x - y)$ we get $< f(x) - f(z), x - z >= \frac{(1-\lambda)}{\lambda} < f(x) - f(z), z - y >\leq 0$. Hence $< f(x) - f(z), x - z >= 0$ and by (5.41) we get $f(x) = f(z)$. Analogously we show $f(y) = f(z)$, so $f(x) = f(y)$, and (5.42) follows. ∎

A generalized network $\mathcal{G} = (I, A, b, C, f)$ is said to be *strictly cyclically monotone* if f if a strictly cyclically monotone function.

Theorem 5.4. [GM3] Let $\mathcal{G} = (I, A, b, C, f)$ be a strictly Cyclically Monotone network with symmetric matrix of connections A. Then any finite orbit $(x(t)) : t \in \mathbb{Z}_T)$ has period $T \leq 2$.

Proof. We have $\Delta_t E_{\mathcal{G}} = 0$ along the orbit. By (5.40) we deduce:
$g(u(t)) = g(u(t-2)) + < f(u(t-2)), u(t-2) - u(t) >$ where $u(t) = Ax(t) - b$.
By the preceeding lemma we conclude: $f(u(t-2)) = f(u(t))$ so $x(t-1) = x(t+1)$.
Hence $T \leq 2$. ■

We can characterize the cyclically and strictly cyclically monotone real functions.

Proposition 5.3. $f : \mathbb{R} \to \mathbb{R}$ is cyclically monotone iff it is increasing. In this case f is also strictly cyclically monotone.

Proof. The equality $(f(x) - f(y))(x - y) = 0$ implies $f(x) = f(y)$, so in the real case cyclically monotone functions are also strictly cyclically monotone.

On the other hand f cyclically monotone implies $(f(x) - f(y))(x - y) \geq 0$ so f is necessarily increasing.

Now suppose that f is increasing, let us show that inequality (5.35) holds for any $m \geq 2$ and any cycle $(y(s) : s \in \mathbb{Z}_m)$. We shall do it by recurrence on m. For $m = 2$ we call $y(0) = x$, $y(1) = y$ (so $y(2) = x$) and (5.35) is equivalent to the increasing property. Assume that (5.35) holds for $r \leq m - 1$, we shall prove it holds for m.

Let $(y(s) : s \in \mathbb{Z}_m)$ be a cycle. Recall that property (5.35) is equivalent to
$\sum_{s=0}^{m-1} < f(y(s)), y(s) - y(s+1) >\geq 0$. When $y(0) \leq ... \leq y(m-1)$ we get

$\sum_{s=0}^{m-2} f(y(s))(y(s) - y(s+1)) \geq f(y(0))(y(0) - y(m-1))$. Then $\sum_{s=0}^{m-1} f(y(s))(y(s+1))$
$\geq f(y(0))(y(0) - y(m-1)) + f(y(m-1))(y(m-1) - y(0)) =$
$(f(y(m-1)) - f(y(0)))(y(m-1) - y(0)) \geq 0$ so (5.35) holds. The case
$y(0) \geq ... \geq y(m-1)$ is analogous.

Then assume there exists $0 < s_0 < m-1$ such that $y(s_0-1) \geq y(s) \leq y(s_0+1)$
(the other case $y(s_0 - 1) \leq y(s) \geq y(s_0 + 1)$ is analogous) with at least one of the two inequalities strict. We obtain:

$$f(y(s_0 - 1))(y(s_0 - 1) - y(s_0)) + f(y(s_0))(y(s_0) - y(s_0 + 1))$$
$$\geq f(y(s_0 - 1))(y(s_0 - 1) - y(s_0 + 1))$$

Hence $\sum_{s=0}^{m-1} f(y(s))(y(s) - y(s+1)) \geq \sum_{s=0}^{m-2} f(\hat{y}(s))(\hat{y}(s) - \hat{y}(s+1))$ where $\hat{y}(s) = y(s)$ if $0 \leq s \leq s_0 - 1$, $\hat{y}(s) = y(s+1)$ if $s_0 \geq s \geq m-2$, and $\hat{y}(m-1) = y(0)$. By recurrence hypothesis the last sum is non-negative, so the result holds. ∎

To write an explicit formula for the Lyapunov functional associated to an increasing real function f we shall calculate the convex potential g associated to f. For any $x \in \mathbb{R}$ write $\Delta_f^-(x) = f(x) = \lim_{y \to x^-} f(y)$, $\Delta_f^+(x) = \lim_{y \to x^+} f(y) - f(x)$ and $\Delta f(x) = \Delta_f^-(x) + \Delta_f^+(x)$. Hence $D_f = \{x \in \mathbb{R} : \Delta_f(x) > 0\}$ is the countable set of discontinuity points of f. Fix $\alpha \in \mathbb{R}$ and define the discrete part of f starting from α as:

$$f^{d,\alpha}(x) = \Delta_f^+(\alpha) + \sum_{y \in (\alpha, x) \cup D_f} \Delta(y) + \Delta_f^-(x) \quad \text{if } x > \alpha$$

$$f^{d,\alpha}(x) = \Delta_f^-(\alpha) + \sum_{y \in (x,\alpha) \cup D_f} \Delta(y) + \Delta_f^+(x) \quad \text{if } x < \alpha$$

$$f^{d,\alpha}(\alpha) = 0$$

Now the continuity part of f is:

$$f^{c,\alpha} = f - f^{d,\alpha}$$

Obviously $f^{c,\alpha}$ is continuous and $f^{c,\alpha}(\alpha) = f(\alpha)$.

It is easily shown that the following function g is a convex potential associated to f:

$$g(x) = \int_\alpha^x f^{c,\alpha}(u)du + (x - \alpha)f^{d,\alpha}(x) \qquad x \in \mathbb{R} \qquad (5.43)$$

From (5.37) the polar function satisfies:

$$g^*(f(x)) = xf(x) - \int_\alpha^x f^{c,\alpha}(u)du - (x - \alpha)f^{d,\alpha}(x) \qquad (5.44)$$

In the multidimensional case we get:

Proposition 5.4. Let I be finite and $f_i : \mathbb{R} \to \mathbb{R}$ be increasing for any $i \in I$. Then the function $f = (f_i : i \in I)$ is strictly cyclically monotone. Pick $\alpha = (\alpha_i : i \in I) \in \mathbb{R}^{|I|}$ then:

$$g(x) = \sum_{i \in I} \left(\int_{\alpha_i}^{x_i} f_i^{c,\alpha_i}(u)du + (x_i - \alpha_i)f_i^{d,\alpha_i}(x) \right) \qquad (5.45)$$

is a convex potential associated to f. The polar g^* related to g satisfies:

$$g^*(f(x)) = \sum_{i \in I} \left(x_i f_i(x_i) - \int_{\alpha_i}^{x_i} f_i^{c,\alpha_i}(u) du - (x_i - \alpha_i) f^{d,\alpha_i}(x_i) \right) \qquad (5.46)$$

Proof. It is direct from Proposition 5.3 and equalities (5.43), (5.44). ∎

From Theorems 5.3, 5.4 and last Proposition we obtain:

Theorem 5.5. [FGM,GM3] Let I be finite and assume for any $i \in I$ the function $f_i : \mathbb{R} \to \mathbb{R}$ to be continuous and increasing. Let $f(x) = (f_i(x_i) : i \in I)$ be the function from $\mathbb{R}^n$ into $C = \mathbb{R}^n$.

For a symmetric connection matrix A consider $\mathcal{G} = (I, A, b, C, f)$ and $(x(t) : t \geq 0)$ an orbit of it. Fix α_i for $i \in I$. Then the functional:

$$E_\mathcal{G}(x(t)) = \langle x(t), Ax(t-1) \rangle$$

$$- \sum_{i \in I} \left(\int_{\alpha_i}^{\sum_{j \in I} a_{ij} x_j (t-1) - b_i} f_i(u) du + \int_{\alpha_i}^{\sum_{j \in I} a_{ij} x_j (t) - b_i} f_i(u) du \right) \qquad (5.47)$$

is a Lyapunov one.

Furthermore, if $(x(t) : t \in \mathbb{Z}_T)$ is a periodic orbit of $\mathcal{G}$ then its period satisfies $T \leq 2$.

Proof. To prove that $E_\mathcal{G}(t)$ is decreasing we apply directly (5.39) and (5.45) by taking into account that $f_i^{d,\alpha_i} = 0$. From Theorem 5.4 and Proposition 5.4 we deduce $T \leq 2$. ∎

Recall that functional (5.47) is the same as that obtained in (5.33) by the approximation method, it is only shifted in one step of time.

5.6. Positive Definite Interactions. The Maximization Problem

For generalized automata we have been able to show that limit cycles are of period $T \leq 2$. For some special cases we can gain in accuracy. To this purpose recall that an $n \times n$ symmetric matrix A is *semi-positive definite* if:

$$\langle x, Ax \rangle \geq 0 \text{ for any } x \in \mathbb{R}^n \qquad (5.48)$$

It is *positive definite* if it also satisfies:

$$< x, Ax >= 0 \text{ iff } x = 0 \tag{5.49}$$

A symmetric matrix is called *semi-negative definite* if $< x, Ax > \leq 0$ for $x \in I\!\!R^n$ and *negative definite* if it also satisfies: $< x, Ax >= 0$ iff $x = 0$. For this kind of matrices we have.

Proposition 5.5. [GM3] Let $\mathcal{G} = (I, A, b, C, f)$ be a strictly cyclically monotone automata network with symmetric matrix of connections A. Then:
   (i) if A is positive definite any periodic orbit is a fixed point,
   (ii) if A is negative definite, $0 \in C$ and $f(0) = 0$, the only fixed point is $x = 0$. Any other periodic orbit is of exact period $T = 2$.

Proof. Denote by g the convex potential associated to f. From Theorem 5.4 any periodic orbit is of period $T \leq 2$. So in the first case we must prove that $T = 2$ is excluded and in the second one that $T = 1$ is excluded except for $x = 0$.
   (i) Let $(x(t) : t \in Z\!\!\!Z_2)$ be a period-2 orbit. By hypothesis the quantity $\gamma =< x(0) - x(1), A(x(0) - x(1)) >$ is ≥ 0. We can write

$$\gamma =< x(0), A(x(0) - x(1)) > + < x(1), A(x(1) - x(0)) >$$

As $x(t + 1) = x(t - 1)$ we deduce:

$$< x(t), A(x(t) - x(t+1)) >=< f(Ax(t-1) - b), (Ax(t) - b) - A(x(t-1) - b) >$$

$\leq g(A(x(t) - b) - g(Ax(t-1) - b)$ where g is the convex potential associated to f. Hence $\gamma \leq g(Ax(0) - b) - g(Ax(1) - b) + g(Ax(1) - b) - g(Ax(0) - b) = 0$. Then $\gamma = 0$. Condition (5.49) implies $x(1) = x(0)$, hence $T = 1$.
   (ii) Assume that $x(0)$ is a fixed point i.e. $x(0) = f(Ax(0))$. We have:

$$g(0) - g(Ax(0)) \geq < f(Ax(0)), -Ax(0) >=< x(0), -Ax(0) >\geq 0$$

The latter inequality rests on A being negative definite. Now, since $f(0) = 0$ and g is a convex potential of f we have $g(y) \geq g(0)$ for any $y \in I\!\!R^n$. We conclude $< x(0), Ax(0) >= 0$, so $x(0) = 0$. ∎

If A is semi-positive, simple Lyapunov functionals can be constructed for positive networks. To this purpose denote $\|x\|_A^2 =< x, Ax >$ for $x \in I\!\!R^n$.

Proposition 5.6. [GM1] Assume A to be symmetric positive definite. Then $E^*(t) = -\|x(t)\|_A^2$ is a Lyapunov functional for the positive network $\mathcal{G} = (I, A, b = 0, C, f)$.

Proof. Let $x(t) = f(Ax(t-1))$ be a trajectory. Since f is positive we deduce:

$$< x(t) - x(t-1), Ax(t-1) > = < f(Ax(t-1)) - f(Ax(t-2)), Ax(t-1) > \geq 0.$$

Then $\phi_f(Ax(t-1)) = < x(t), Ax(t-1) > \geq < x(t-1), Ax(t-1) > = \|x(t-1\|_A^2$.
    Since A is semi-positive definite:

$$0 \leq < x(t) - x(t-1), A(x(t) - x(t-1)) > = \|x(t)\|_A^2 + \|x(t-1)\|_A^2 - 2\phi_f(Ax(t-1)).$$

Hence necessarily $\|x(t)\|_A^2 \geq \phi_f(Ax(t-1))$.
    Since $b = 0$ the function $-\phi_f(Ax(t-1))$ is a Lyapunov functional. From the inequalities $-\|x(t)\|_A^2 \leq -\phi_f(Ax(t-1)) \leq -\|x(t-1)\|_A^2$ we conclude the result.
∎

Corollary 5.6. Let $g : D \to I\!\!R$ be a positive homogeneous convex function (for instance g a seminorm) and $f : D \to C$ be a subgradient of it. Assume A to be symmetric positive definite. Then any solution of one of the following maximization problems:

$$g(Ay_0) = \max_{y \in C} g(Ay), \quad y_0 \in C \tag{5.50}$$

$$\|z_0\|_A^2 = \max_{z \in C} \|z\|_A^2, \quad z_0 \in C \tag{5.51}$$

is a fixed point of the positive automaton $\mathcal{G} = (I, A, b = 0, C, f)$.

Proof. It is deduced from Corollary 5.3 and Propositions 5.5, 5.6. ∎

5.7. Sequential Iteration for Decreasing Real Functions and Optimization Problems

    Let $f : I\!\!R \to I\!\!R$ be an increasing function such that $f(0) = 0$. Assume that A is a symmetric matrix with $\text{diag} A = 0$ and the threshold vector $b = 0$. The sequential updating of this continuous networks is:

$$x_i' = f\left(\sum_{j \in I} a_{ij} x_j\right)$$

As with the similar results obtained in the previous and present chapter, here we have:

Proposition 5.7. [FGM] Let $\{x(t)\}_{t \geq 0}$ be a sequential trajectory. The quantity:

$$E_{seq}(x(t)) = -\frac{1}{2} < x(t), Ax(t) > + \sum_{i \in I} \int_0^{x_i(t)} f^{-1}(\varsigma)d\varsigma \qquad (5.52)$$

is a Lyapunov functional.

Proof. Let i be the site to update. From symmetry of A and $\text{diag}A = 0$, we get:

$$\Delta_t E_{seq} = -(x_i(t+1) - x_i(t)) \sum_{j \in I} a_{ij}x_j(t) + \int_0^{x_i(t+1)} f^{-1}(\varsigma)d\varsigma - \int_0^{x_i(t)} f^{-1}(\varsigma)d\varsigma$$

Denote $u = \sum_{j \in I} a_{ij}x_j(t)$, $v = \sum_{j \in I} a_{ij}x_j(t-1)$. The above equality is written:

$$\Delta_t E_{seq} = -(f(u) - f(v))u + \int_0^{f(u)} f^{-1}(\varsigma)d\varsigma - \int_0^{f(v)} f^{-1}(\varsigma)d\varsigma$$

Since

$$\int_0^{f(x)} f^{-1}(\varsigma)d\varsigma = xf(x) - \int_0^x f(\varsigma)d\varsigma$$

we deduce:

$$\Delta_t E_{seq} = f(v)(u - v) + \int_0^v f(\varsigma)d\varsigma - \int_0^u f(\varsigma)d\varsigma \leq 0$$

which is non positive because f is increasing. ■

Remarks.
1. It is straighforward to extend to above result when there exists a threshold vector $b \in I\!R^n$ and $\text{diag}A \geq 0$. In fact in this situation one obtains as Lyapunov functional previous expression (5.52) plus the linear term $< b, x(t) >$ [FGM].
2. The usual real functions f taken in optimization are of the form f strictly increasing in an interval $[-s, s]$ and constant elsewhere, i.e. equal to $f(-s)$ if $x \leq -s$, $f(s)$ if $x \geq s$. In this case one also obtains Lyapunov functional (5.52), but the proof takes into account more cases [FGM]. Also, the assumption $f(0) = 0$

is not important; it is always possible to change the initial point of the integral to another value $\alpha \in \mathbb{R}$.

In such framework sequential iteration has been used as a hill-climbing strategy for some hard combinatorial optimization problems as the Travelling Salesman Problem (TSP) [HoT]. The idea is to approach the set of solution of TSP by the minimization of a quadratic gain functional with a symmetric matrix A and $\mathrm{diag}\, A = 0$. By doing so, the sequential update with threshold or sign functions is associated to the quadatic cost function (as we did in Proposition 3.8) and its fixed points are local minima of the optimization problem. Furthermore, the overall minima are also fixed points of the sequential update.

Let us explicit this methodology. Consider the combinatorial optimization problem:

$$\min_{x \in \{-1,1\}^n} E(x) = -\frac{1}{2} < x, Ax >, \tag{5.53}$$

where A symmetric, $\mathrm{diag}\, A = 0$.

Problem (5.53) is in general hard to solve. For instance, computing ground states of three-dimensional spin glasses is a particular case of (5.53) and it is NP-complete [Ba].

It is direct that the sequential iteration:

$$x'_i = \mathrm{sign}\left(\sum_{j \in I} a_{ij} x_j\right)$$

admits (5.53) as Lyapunov functional (Proposition 3.8), hence its fixed points are local minima of (5.53). Unfortunately the fixed points obtained are high local minima, hence bad solutions. To avoid this problem one associates to the sequential update a bounded increasing function f_T which converges pointwise to sign when $T \to 0$. From previous developments one knows that:

$$E_T\left(x(t)\right) = -\frac{1}{2} < x(t), Ax(t) > + \sum_{i \in I} \int_0^{x_i(t)} f_T^{-1}(\varsigma)d\varsigma \tag{5.54}$$

is a Lyapunov functional for the sequential update.

Furthermore, as $T \to 0$, $E_T \to E$, hence (5.54) is a perturbation of the problem (5.53). In practice, for good choices of the parameter T (the temperature) the functional E_T has a smoother landscape. Hence the convergence to stationary points which are high local minima decreases. Two functions usually used are the sigmoid and the "Brain State in the Box":

$$\mathrm{sgm}_T(u) = \begin{cases} -1 & \text{if } u \leq -T \\ \mathrm{tgh}\frac{u}{T} & \text{if } -T \leq u \leq T \\ +1 & \text{if } u \geq T \end{cases}$$

$$BSB_T\left(u\right) = \begin{cases} -1 & \text{if } u \le -T \\ \frac{u}{T} & \text{if } -T \le u \le T \\ +1 & \text{if } u \ge T \end{cases}$$

Such continuous strategies have been used for several applications, see for instance [Ho1,HoT,PA1,PA2]. In [PA1,PA2] a theoretical approach in the framework of the Mean Field Theory was given for the sigmoid function. To determine the best choice for the parameter T, a possibility is to use a deterministic annealing schedule. That is to say, in the iteration scheme, the temperature decreases slowly to zero which is equivalent to change slowly the shape of the iteration function until reaching the sign rule [GHM,KGV,PA1].

Besides, in practice simulations of sigmoid and BSB models lead to very similar solutions of the discrete optimization problem (5.51). Such situation is illustrated by the graph, due to Mejia, in Figure 5.3, where a set of ten random 40x40 symmetric matrices with $a_{ij} \in [-1, +1]$ and $\text{diag} A = 0$ was used for the sgm, BSB and sign rules in sequential update.

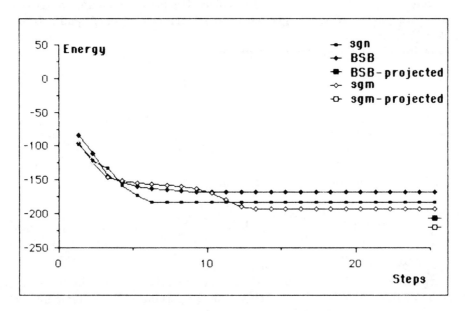

Figure 5.3. Mean values of cost functions for the sequential update of sgm, BSB and sign rules. The stationary vectors obtained for BSB and sgm are projected in the hypercube: $p(x) = \text{sign}(x)$.

5.8. A Generalized Dynamics

For applications which we shall make in the problem of describing Gibbs states of Bethe lattice the following framework will be useful. Let $D \subset \mathbb{R}^n$ be a closed convex set satisfying $A(D) \subset D$. The set of configurations is a subset $C \subset \mathbb{R}^n$.

For $\varphi : C \to D$ and $h : D \to C$ consider the dynamics from C into C:

$$x(t+1) = F(x(t)) \text{ where } F = h \circ A \circ \varphi \qquad (5.55)$$

This means $x(t+1) = h(A(\varphi(x(t))))$.

Denote $f = \varphi \circ h$ the funcion mapping D into D. We have:
$x(t) = h \circ A \circ ((f \circ A)^{t-1}(\varphi(x(0))))$, or equivalently:

$$\varphi(x(t)) = (f \circ A)^t (\varphi(x(0))) \qquad (5.56)$$

Hence the above dynamics can be written:

$$y(t+1) = f(Ay(t)) \text{ where } y(t) = \varphi(x(t)) \qquad (5.57)$$

So all the information we have for the class of automata $\mathcal{G} = (I, A, b = 0, D, f)$ can be strictly used in the study of dynamics (5.49).

Proposition 5.8. [GM4] Consider the dynamics $x(t+1) = h \circ A \circ \varphi(x(t))$. Assume that $f = \varphi \circ h$ is cyclically monotone, let g be its potential. Then:

$$E(x(t)) = <\varphi(x(t-1)), A\varphi(x(t))> -g(A\varphi(x(t-1))) - g(A\varphi(x(t)))$$

is a Lyapunov functional. If f is also strictly cyclically monotone then the orbits of the dynamics are of period $T \leq 2$.

If A is strictly positive definite the orbits are only fixed points. When A is strictly negative definite, $0 \in D$ and $f(0) = 0$ then $h(0)$ is the only fixed point, any other orbit being of period $T = 2$.

Proof. From expression (5.57), Theorems 5.3, 5.4. and Proposition 5.5. ∎

If f is positive the Lyapunov functional becomes:

$$E(x(t)) = -\phi_f(A\varphi(x(t))) \text{ where } \phi_f(x) = <f(x), x> \qquad (5.58)$$

Example 5.4. In the next chapter we shall deal with evolutions of type (5.55). In that case $A \geq 0$ (this means $a_{ij} \geq 0$ for any $i, j \in I$),

$\mathcal{D} = I\!\!R^n_+ = \{x \in I\!\!R^n : x_i \geq 0 \text{ for any } i \in I\}$, so the hypothesis $A(\mathcal{D}) \subset \mathcal{D}$ is satisfied. Take $p > 0$. The set of configurations is $C = \{x \in I\!\!R^n_+ : \|x\|_{p+1} = 1 \text{ or } x = 0\}$ and we consider:

$$\varphi(x) = x^p = (x^p_i : i \in I), \quad h(x) = \begin{cases} \frac{x}{\|x\|_{p+1}} & \text{if } x \neq 0 \\ 0 & \text{if } x = 0 \end{cases}$$

Then $f(x) = \varphi \circ h(x) = \frac{x^p}{\|x\|^p_{p+1}}$ is a strictly cyclically monotone function. In fact f is example 5.3 of section 5.2 and we showed there that it was a strictly positive function. For this case $\phi_f(x) = \|x\|_{p+1}$ so (5.58) takes the form:

$$E(x(t)) = -\|A\varphi(x(t))\|_{p+1} \tag{5.59}$$

5.9. Chain-Symmetric Matrices

In this chapter we have introduced the convex analysis techniques which allow the definition of a Lyapunov functional and the conclusion that the orbits are necessarily of period $T = 1$ or 2.

Let us read in a new way the hypotheses we have imposed.

Consider the evolution:

$$x(t+1) = f(Ax(t) - b), \quad x(t) \in C \subset I\!\!R^{|I|} \tag{5.60}$$

We needed f cyclically monotone, which means that there exists a function (the potential) $g : \mathcal{D} \to I\!\!R$ with $\mathcal{D} \subset I\!\!R^{|I|}$ a convex set, satisfying:

$$g(x) \geq g(y) + < f(y), x - y > \text{ for any } x, y \in \mathcal{D} \tag{5.61}$$

We also assumed A to be symmetric, i.e.:

$$< Ax, y > = < x, Ay > \text{ for any } x, y \in \mathcal{D} \tag{5.62}$$

Hypotheses (5.61) and (5.62) implied that the functional defined in (5.39) was a Lyapunov one. Furthermore if f also satisfied the strict property:

$$< f(x) - f(y), x - y > = 0 \text{ iff } f(x) = f(y) \tag{5.63}$$

then we could conclude that the orbits of evolution (5.60) were of period $T = 1$ or 2.

Now recall that conditions (5.61), (5.62) and (5.63) depend on the inner product $<,>$. Hence all the results must be read with the predicate: for some inner product $<,>$. Obviously this inner product must be the same for the properties (5.61), (5.62) and (5.63). If condition (5.62) holds we say A is $<,>$ self adjoint.

It can be shown (see [CP]) that the class of matrices which are self-adjoint for some inner product are the diagonalizable ones. But a problem appears, in fact the cyclically monotone property (or the positive property (5.6)) depends on the inner product. This means a function f can be cyclically monotone (or positive) for some inner product but it can fail to have this property for a different one.

Then if, at a first glance, we can extend our results for diagonalizable matrices we can only accept for a certain matrix A the functions f which are cyclically monotone for the inner product which makes A self-adjoint.

But there exist some inner products which preserve a big class of usual cyclically monotone functions. In particular the threshold and the sign function continue to be cyclically monotone for such inner products. We shall introduce the class of matrices related to these inner products.

Definition 5.3. A matrix $A = (a_{ij} : i, j \in I)$ is called *chain symmetric* iff the following two conditions hold:

$$\text{sign}(a_{ij}) = \text{sign}(a_{ji}) \quad \text{for any } i, j \in I \tag{5.64}$$

$$a_{i_0 i_1} a_{i_1 i_2} \dots a_{i_{k-2}, i_{k-1}} a_{i_{k-1}, i_0} = a_{i_0, i_{k-1}} a_{i_{k-1}, i_{k-2}} \dots a_{i_2 i_1} a_{i_1 i_0} \tag{5.65}$$

for any sequence $i_0 \dots i_{k-1} \in I$.

Obviously a symmetric matrix is chain symmetric. On the other hand, there exist chain symmetric matrices which are not symmetric. An example of one of these matrices is:

$$A = \begin{pmatrix} 1 & 2 & 6 \\ 1 & 1 & 3 \\ 1 & 1 & 1 \end{pmatrix}$$

Now recall that any strictly positive vector $\mu = (\mu_i > 0 : i \in I)$ defines an inner product on $\mathbb{R}^{|I|}$:

$$< x, y >_\mu = \sum_{i \in I} \mu_i x_i y_i \tag{5.66}$$

A matrix A is $<>_\mu$ self-adjoint if $< Ax, y >_\mu = < x, Ay >_\mu$ for any $x, y \in \mathbb{R}^{|I|}$ or equivalently:

$$\mu_i a_{ij} = \mu_j a_{ji} \quad \text{for any } i, j \in I \tag{5.67}$$

We have the following characterization of chain symmetry:

Proposition 5.9. [M4] $A = (a_{ij} : i \in I)$ is chain symmetric iff there exists some strictly positive vector $\mu > 0$ such that A is $<>_\mu$ self-adjoint.

Proof. First assume that there exists some $\mu > 0$ such that A is $<>_\mu$ self-adjoint. From the equalities $\mu_i a_{ij} = \mu_j a_{ji}$, we get $\text{sign}(a_{ij}) = \text{sign}(a_{ij})$. Now consider $i_0, ..., i_{k-1} \in I$ and denote $i_k = i_0$. If for some ℓ we have $a_{i_\ell, i_{\ell+1}} = 0$ then $a_{i_{\ell+1}, i_\ell} = 0$ so equality (5.65) follows. Then suppose $a_{i_\ell, i_{\ell+1}} \neq 0$ for $\ell = 0, ..., k-1$. Hence

$$\frac{a_{i_\ell, i_{\ell+1}}}{a_{i_{\ell+1}, i_\ell}} = \frac{\mu_{i_\ell}}{\mu_{i_{\ell+1}}} \frac{\mu_{i_{\ell+1}}}{\mu_{i_\ell}} = 1$$

so (5.65) is obtained.

Now assume that A is chain symmetric. Let us construct some $\mu > 0$ satisfying $\mu_i a_{ij} = \mu_j a_{ji}$. Define the relation $i \overset{0}{\to} i$, $i \overset{1}{\to} j$ if $j \neq i$ and $a_{ij} \neq 0$. For $k > 1$ we write $i \overset{k}{\to} j$ iff $j \neq i$ and there exists a sequence $i_1, \cdots, i_{k-1} \in I$ such that $i \overset{1}{\to} i_1 \overset{1}{\to} \cdots i_{n-2} \overset{1}{\to} i_{n-1} \overset{1}{\to} j$, with $k-1$ being the minimal cardinality of a sequence satisfying the above property.

Fix $i_0 \in I$ and define $\mu_{i_0} > 0$. For any $j \in I$ such that $i \overset{1}{\to} j$ define $\mu_j = \frac{\mu_{i_0} a_{i_0 j}}{a_{j i_0}}$. Now suppose μ is well defined on the set

$$\{j \in I : i_0 \overset{\ell}{\to} j \quad \text{for } \ell \leq k-1\},$$

let us define it for $j \in I$ such that $i_0 \overset{k}{\to} j$. Let $i_0 \overset{1}{\to} i_1 \overset{1}{\to} \cdots \overset{1}{\to} i_{k-1} \overset{1}{\to} j$. By the minimality of k we have $i_0 \overset{\ell}{\to} i_\ell$ for $\ell = 1, \cdots, k-1$. We define:

$$\mu_j = \frac{\mu_{i_{k-1}} a_{i_{k-1}, j}}{a_{j, i_{k-1}}}.$$

We shall prove that μ_j is well defined, i.e. if $i_0 \overset{1}{\to} j_1 \overset{1}{\to} \cdots \overset{1}{\to} j_{k-1} \overset{1}{\to} j$ we must show:

$$\frac{\mu_{i_{k-1}} a_{i_{k-1}, j}}{a_{j, i_{k-1}}} = \frac{\mu_{j_{k-1}} a_{j_{k-1}, j}}{a_{j, j_{k-1}}} \tag{5.68}$$

Now we have:

$$\frac{\mu_{i_{k-1}}}{\mu_{i_0}} = \frac{\mu_{i_{k-1}}}{\mu_{i_{k-2}}} \cdots \frac{\mu_{i_2}}{\mu_{i_1}} \frac{\mu_{i_1}}{\mu_{i_0}} = \frac{a_{i_{k-2}, i_{k-1}}}{a_{i_{k-1}, i_{k-2}}} \cdots \frac{a_{i_1 i_2}}{a_{i_2 i_1}} \frac{a_{i_0 i_1}}{a_{i_1 i_0}}$$

Analogously

$$\frac{\mu_{j_{k-1}}}{\mu_{i_0}} = \frac{a_{j_{k-2}, j_{k-1}}}{a_{j_{k-1}, j_{k-2}}} \cdots \frac{a_{j_1 j_2}}{a_{j_2 j_1}} \frac{a_{i_0 j_1}}{a_{j_1 i_0}}.$$

Then equality (5.68) is equivalent to:

$$a_{i_{k-2},i_{k-1}} \cdots a_{i_1 i_2} a_{i_0 i_1} a_{j_{k-1},j_{k-2}} \cdots a_{j_2 j_1} a_{j_1 i_0} a_{i_{k-1},j} a_{j,j_{k-1}} =$$
$$a_{j_{k-2},j_{k-1}} \cdots a_{j_1 j_2} a_{i_0 j_1} a_{i_{k-1},i_{k-2}} \cdots a_{i_2 i_1} a_{i_1 i_0} a_{j_{k-1},j} a_{j,i_{k-1}}$$

By arranging terms the above equality is written:

$$a_{j,j_{k-1}} a_{j_{k-1},j_{k-2}} \cdots a_{j_2 j_1} a_{j_1 i_0} a_{i_0 i_1} a_{i_1 i_2} \cdots a_{i_{k-2},i_{k-1}} a_{i_{k-1},j} =$$
$$a_{j,i_{k-1}} a_{i_{k-1},i_{k-2}} \cdots a_{i_2 i_1} a_{i_1 i_0} a_{i_0 j_1} a_{j_1 j_2} \cdots a_{j_{k-2},j_{k-1}} a_{j_{k-1},j}$$

But this is exactly condition (5.68). Then μ is well defined for all j such that $i \overset{k}{\to} j$ iff $i \overset{k}{\to} j$ for some $k \geq 0$.

Now the relation $i \overset{*}{\to} j$ iff $i \overset{k}{\to} j$ for some $k \geq 0$, is an equivalence relation (the symmetry of $\overset{*}{\to}$ is deduced from (5.64)). Then we define μ by fixing some element i_0 in each class and making the above construction.

Let us show that $\mu_i a_{ij} = \mu_j a_{ji}$ holds for any couple $i \neq j$ in I. If $i \overset{1}{\to} j$ is not satisfied then $a_{ij} = a_{ji} = 0$, hence the above equality is satisfied. Then assume $i \overset{1}{\to} j$, which implies that they belong to the same class; call i_0 the element of this class from which we have constructed μ. If $i_0 \overset{k}{\to} i$, $i_0 \overset{k'}{\to} j$ we have $|k - k'| \leq 1$.

If $k = k'$, let $i_0 \overset{1}{\to} i_1 \overset{1}{\to} \cdots \overset{1}{\to} i_{k-1} \overset{1}{\to} i$, $i_0 \overset{1}{\to} j_1 \overset{1}{\to} \cdots \overset{1}{\to} j_{k-1} \overset{1}{\to} j$. We have

$$\frac{\mu_i a_{ij}}{\mu_j a_{ji}} = \frac{a_{ij}}{a_{ji}} \frac{\mu_i}{\mu_{i_{k-1}}} \frac{\mu_{i_{k-1}}}{\mu_{i_{k-2}}} \cdots \frac{\mu_{i_2}}{\mu_{i_1}} \frac{\mu_{i_1}}{\mu_{i_0}} \frac{\mu_{i_0}}{\mu_{j_1}} \frac{\mu_{j_1}}{\mu_{j_2}} \cdots \frac{\mu_{j_{k-2}}}{\mu_{j_{k-1}}} \frac{\mu_{j_{k-1}}}{\mu_j} =$$

$$\frac{a_{ij}}{a_{ji}} \frac{a_{i_{k-1},i}}{a_{i,i_{k-1}}} \frac{a_{i_{k-2},i_{k-1}}}{a_{i_{k-1},i_{k-2}}} \cdots \frac{a_{i_1 i_2}}{a_{i_2 i_1}} \frac{a_{i_0 i_1}}{a_{i_1 i_0}} \frac{a_{j_1 i_0}}{a_{i_0 j_1}} \frac{a_{j_2 j_1}}{a_{j_1 j_2}} \cdots \frac{a_{j_{k-1},j_{k-2}}}{a_{j_{k-2},j_{k-1}}} \frac{a_{j,j_{k-1}}}{a_{j_{k-1},j}}.$$

From (5.65) we deduce that this expression is equal to 1, hence the required equality. If $k' = k + 1$ there exists a sequence $i \overset{1}{\to} i_1 \overset{1}{\to} \cdots i_{k-1} \overset{1}{\to} i \overset{1}{\to} j$. Since μ is well defined we get $\mu_i a_{ij} = \mu_j a_{ij}$. The case $k = k' + 1$ is analogous. Then the assertion of the Proposition holds. ∎

Now consider $f : \mathbb{R}^{|I|} \to \mathbb{R}^{|I|}$ of the form $f(x) = (f_i(x_i) : i \in I)$ with f_i an increasing real function. It is easy to show that f is $<>_\mu$-strictly cyclically monotone. The proof is evident because it suffices to consider the case $|I| = 1$. Hence in all the results obtained in this chapter for the above class of functions f the hypothesis "A symmetric" can be replaced by the less restrictive one "A chain symmetric".

References

[An] Anderson, J.A., *Cognitive and Phychological Computation with Neural Models*, IEEE Transactions on Systems, Man and Cybernetics, SMC-13, 1983, 799-815.

[Ba] Barahona, F., *Application de l'Optimisation Combinatoire à Certains Modeles de Verres de Spin*, Complexite et Simulations, Tesis, IMAG, Grenoble, France, 1980.

[CP] Cailliez, F., J.P. Pages, *Introduction a l'Analyse des Données*, Smash, 1976.

[FGM] Fogelman-Soulié, F., E. Goles, S. Martínez, C. Mejía, *Energy Functions in Neural Networks with Continuous Local Functions* 1988, Submitted to Complex Systems.

[G1] Goles, E., *Dynamics on Positive Automata*, Theoret. Comput. Sci., 41, 1985, 19-31.

[GHM] Goles, E., G. Hernández, M. Matamala., *Dynamical Neural Schema for Quadratic Discrete Optimization Problems*, Neural Networks, 1, Supplement 1, 1988, 96.

[GM1] Goles, E., S. Martínez, *A Short Proof on the Cyclic Behaviour of Multithreshold Symmetric Automata*, Information and Control, 51(2), 1981, 95-97.

[GM2] Goles, E., S. Martínez, *Properties of Positive Functions and the Dynamics of Associated Automata Networks*, Discreta Appl. Math. 18, 1987, 39-46.

[GM3] Goles, E., S. Martínez, *Lyapunov Functionals for Automata Networks defined by Cyclically Monotone Functions*, Preprint, 1987, Submitted to SIAM Journal on Discrete Maths.

[GM4] Goles, E., S. Martínez, *The One-Site Distributions of Gibbs States on Bethe Lattice are Probability Vectors of Period ≤ 2 for a Nonlinear Transformation*, J. Stat. Physics, 52(1/2), 1988, 267-285.

[GO1] Goles, E., J. Olivos, *Comportement Iteratif des Fonctions à Multiseuil*, Information and Control, 45(3), 1980, 800-813.

[Ho1] Hopfield, J.J., *Neurons with Graded Response have Collective Computational Properties like those of two-state Neurons*, Proc. Nat. Acad, Sci, USA, 81, 1984, 3088-3092.

[HoT] Hopfield, J.J., D.W. Tank, *Neural Computation of Decisions in Optimization Problems*, Biol. Cybernetics, 52, 1985, 141-152.

[KGV] Kirkpatrick, S., C. Gelatt, M. Vecchi, *Optimization by Simulated Annealing*, Science, 220, 1983, 671-680.

[M3] Martínez, S., *Relations among Discrete and Continuous Lyapunov Functionals for Automata Networks*, Preprint, 1989.

[M4] Martínez, S., *Chain-Symmetric Automata Networks*, Preprint, 1989.

[PA1] Peterson, C., J. Anderson, *A Mean Field Theory Learning Algorithm for Neural Networks*, Complex System, 1(5), 1987, 995-1019.

[PA2] Peterson, C., J. Anderson, *Neural Networks and NP-Complete Optimization Problems; A Performance Study on the Graph Bisection Problem*, Complex System, 2(1), 1988, 59-89.

[PE] Pham Dinh Tao, S. El Bernoussi, *Iterative Behaviour, Fixed Point of a Class of Monotone Operators. Application to Non-Sysmmetric Threshold Functions*, Disc. Maths., 70, 1988, 85-101.

[PT1] Poljak, S., D. Turzik, *On Pre-Periods of Discrete Influence Systems*, Disc. Appl. Maths, 13, 1986, 33-39.

[PT2] Poljak, S., D. Turzik, *On an Application of Convexity to Discrete Systems*, Disc. Appl. Math. 13, 1986, 27-32.

[Rck] Rockafellar, R.T., *Convex Analysis*, Princeton Univ. Press, Princeton, NJ., 1970.

6. APPLICATIONS ON THERMODYNAMIC LIMITS ON THE BETHE LATTICE

6.1. Introduction

In this chapter we will use the general Lyapunov functionals introduced in Chapter 5 to study and describe the thermodynamic limit of Gibbs ensembles in Bethe lattice L_∞.

We have proved that these thermodynamic limits satisfy evolutions which can be put in the form:

$$x(t+1) = h \circ A \circ \varphi(x(t))$$

where A is a self-adjoint matrix with respect to some inner product $< >$ and $f = \varphi \circ h$ is a cyclically monotone function with respect to $< >$ of some convex potential which, in this case, turns out to be a norm. Hence by the results of sections 5.8, 5.9 we are able to conclude that the limit points x^* of the above system satisfy a period-2 equation (see Theorems 6.1, 6.2):

$$x^* = U^2 x^* \text{ where } U = h \circ A \circ \varphi.$$

The above results were first shown in [GM4,M1,M2].

In the case $L_\infty = \mathbb{Z}$, the integer lattice, the above equation reduces to a Perron-Frobenius equation (see Corollary 6.1), and this was first established in reference [Br]. Deeper results concerning characterization of Gibbs states on Bethe lattice by using entropy arguments can be found in [FV].

We recall that the definitions of Gibbs ensemble and thermodynamic limits, as well as the general theoretical framework of Statistical Physics, can be found in reference [Ru].

In section 6.6 we will study some non linear evolution of probability vectors. In particular we will describe the limit behaviour of the dynamics:

$$x_i(t+1) = \frac{\sum\limits_{j \in I} a_{ij} x_j(t)^p}{\sum\limits_{i \in I} \sum\limits_{j \in I} a_{ij} x_j(t)^p} \qquad i \in I$$

for $p > 0$, $A = (a_{ij} : i, j \in I)$ non-negative and chain-symmetric.

Recall that in the case $p = 1$, $\sum\limits_{j \in I} a_{ij} = 1$ for any $i \in I$, and this corresponds to a Markov chain. Then our results describe the limit behaviour of non-linear evolutions which generalyze Markov chains. But our characterization of the limit states x^* is not as rich as is in the linear case $p = 1$ because we are not able to describe the domain of attraction of the solution.

6.2. The Bethe Lattice

Let $q \geq 2$. Consider the graph constructed as follows: we start from a central point 0 and we add q points all connected to 0. Call the set of these new points shell 1. For $t \geq 1$ shell $t+1$ is constructed by connecting $q-1$ new points to any point of shell t. The graph constructed up to shell t is denoted L_t and the infinite graph obtained as $t \to \infty$, denoted L_∞, is called the Bethe Lattice of coordination number q. Recall that for $q = 2$ the Bethe Lattice is $L_\infty = \mathbb{Z}$.

Let us be more precise. Denote $Q_1 = \{1, ..., q\}$, $Q = \{1, ..., q-1\}$. Shell t is formed by the points $l = (l_1, ..., l_t)$ with $l_1 \in Q_1$, $l_s \in Q$ for $2 \leq s \leq t$. The set L_t is formed by 0 and the union of shells s with s varying from 1 to t, i.e.:

$$L_t = \{0\} \cup \bigcup_{s=1}^{t} \{(l_1, ..., l_s) : l_1 \in Q_1, l_r \in Q \text{ for } 2 \leq r \leq s, \text{ for } s = 1, ..., t\}$$

$$\text{and } L_\infty = \bigcup_{t \geq 1} L_t \tag{6.1}$$

The neighbours of the central point 0 are the points of shell 1, whereas two points of consecutive shells $(l_1, ..., l_t)$, $(l'_1, ..., l'_{t+1})$ are neighbours if $l'_1 = l_1, ..., l'_t = l_t$. We denote $(l, l') \in V$ a link of the non-oriented graph, iff the sites l, l' are neighbours.

Now for any $t > s \geq 1$ and $l \in L_{s+1} \setminus L_s$ (i.e. l belongs to shell $s+1$) denote $L^*_{t,s}(l) = \{l' \in L_t : l'_r = l_r \text{ for } r = 1, ..., s+1\}$.

Let $t > s \geq 1$, $t' > s' \geq 1$, $l \in L_{s+1} \setminus L_s$, $l \in L_{s'+1} \setminus L_{s'}$. If $t - s = t' - s'$ the graphs $L^*_{t,s}(l)$, $L^*_{t',s'}(l')$ are equivalent, in fact the one-to-one correspondence $(l_1, ..., l_{s+1}, l_{s+2}, ..., l_r) \to (l'_1, ..., l'_{s'+1}, l_{s+2}, ..., l_t)$ preserves the neighbourhood relation in the graphs. Thus we assimilate these equivalent graphs, so we can write:

$$L^*_{t,s}(l) = L^*_{t-s}(l) = L^*_{t-s} = \{l' \in L_t : l'_r = l_r \text{ for } r = 1, ..., s+1\}$$

Hence for any pair t, s with $t > s \geq 1$ we have:

$$L_t \setminus L_s = \bigcup_{l \in L_{s+1} \setminus L_s} L^*_{t-s}(l) \tag{6.2}$$

where the family $(L^*_{t-s}(l) : l \in L_{s+1} \setminus L_s)$ contains mutually disjoint equivalent copies of a graph L^*_{t-s}.

See Figure 6.1 below:

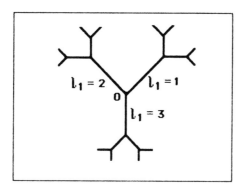

Figure 6.1. Bethe lattice for $q = 3$.

6.3. The Hamiltonian

Let $I \subset \mathbb{R}^d$ be a finite set of spins endowed with the inner product $[\,,\,]$. The set $\Omega_\infty = L_\infty^I$ is the space of (infinite) configurations on L_∞, an element $\sigma \in \Omega_\infty$ is a function $\sigma : L_\infty \to I, l \to \sigma(l)$ which assigns to any node l of the lattice a spin $\sigma(l) \in I$. The space of finite configurations on L_t is denoted by $\Omega_t = L_t^I$ for $t \geq 0$.
For a finite configuration $\sigma \in \Omega_t$ its Hamiltonian is (see reference [Bax]):

$$H_t(\sigma) = -\{K \sum_{(l,l') \in V \cap (L_t \times L_t)} [\sigma(l), \sigma(l')] + \sum_{l \in L_t} [N, \sigma(l)]\} \qquad (6.3)$$

where K is the interaction and $N \in \mathbb{R}^d$ is an exterior magnetic field, the first sum is over all the neighbouring sites in L_t and the second over the sites of L_t.
Note $\Omega_t^*(l) = L_t^*(l)^I$, $\Omega_t^* = L_t^{*I}$. If $\sigma \in \Omega_t$ is a configuration by σ_D we mean the restriction of σ to $D \subset L_t$. Now we write $\sigma^{(l_1)} = \sigma_{L_t^*(l_1)} \in \Omega_t^*(l_1)$ the restriction of σ to $L_t^*(l_1)$. When we develop the Hamiltonian we find:

$$-H_t(\sigma) = <N, \sigma(0)> + \sum_{l_1 \in Q_1} W_t(\sigma(0), \sigma^{(l_1)}) \qquad (6.4)$$

where the function W_t is defined in the following recursive way:

$$W_t(i, \sigma') = [Ki + N, \sigma'(l_1)] + \sum_{l_2 \in Q} W_{t-1}(\sigma'(l_1), \sigma'^{(l_1 l_2)}) \qquad (6.5)$$

$$\text{where } \sigma' \in \Omega_t^*(l_1), \text{ and } \sigma'^{(l_1 l_2)} = \sigma'_{L_{t-1}^*(l_1 l_2)}$$

So if we put $H_s(\sigma) = H_t(\sigma_{L_s})$ for $t > s \geq 1$ we get:

$$-H_t(\sigma) = -H_s(\sigma) + \sum_{l_1 \in Q_1} \sum_{l_2 \in Q} \cdots \sum_{l_{s+1} \in Q} W_{t-s}(\sigma(l_1, ..., l_s), \sigma^{(l_1, ..., l_{s+1})}) \tag{6.6}$$

where $\sigma^{(l_1, ..., l_{s+1})} = \sigma_{L_{t-s}^*}(l_1, ..., l_{s+1})$.

Now define the function:

$$g_t(i) = \sum_{\sigma \in \Omega_t^*} \exp\left(W_t(i, \sigma)\right). \tag{6.7}$$

From (6.2) we deduce that it satisfies the following equations:

$$g_t(i)^q = \sum_{\sigma \in (L_t \backslash \{0\})^I} \exp\left(\sum_{l_1 \in Q_1} W_t(i, \sigma^{(l_1)})\right) \tag{6.8}$$

$$g_t(i) = \sum_{j \in I} g_{t-1}(j))^{q-1} \exp\left([N + Ki, j]\right) \tag{6.9}$$

Assume that the temperature factor $\beta = \frac{1}{kT}$ has been included in the Hamiltonian (we normalize K and N). Then we can express the partition function $Z_t = \sum_{\sigma \in \Omega_t} \exp\left(-H_t(\sigma)\right)$ in terms of $g_t(i)$. In fact by using (6.4), (6.8) we get:

$$Z_t = \sum_{i \in I} (g_t(i))^q \exp\left([N, i]\right) \tag{6.10}$$

To obtain expressions (6.5) through (6.10) we have used the same arguments introduced in reference [Bax], which are near the transfer matrix method used in [Br] to study the case $q = 2$, i.e. $L_\infty = \mathbb{Z}$.

6.4. Thermodynamic Limits of Gibbs Ensembles

Let t be fixed. The Gibbs ensemble ν_t is the following probability measure defined on the set of finite configurations Ω_t:

$$\nu_t(\sigma) = Z_t^{-1} \exp\left(-H_t(\sigma)\right) \text{ for } \sigma \in \Omega_t \tag{6.11}$$

We shall define the thermodynamic limits of the family of Gibbs ensembles $(\nu_t : t \geq 1)$ when $L_t \to L_\infty$.

For $s \leq t \leq \infty$ define the following continuous function (with respect to product topologies on Ω_t, Ω_s):

$$\theta_{s,t} : \Omega_t \to \Omega_s, \text{ such that } \theta_{s,t}\sigma = \sigma_{L_s}, \text{ the restriction of}$$
the configuration $\sigma \in \Omega_t$ to Ω_s, i.e. $(\theta_{s,t}\sigma)(l) = \sigma(l)$ for any $l \in L_s$. $\qquad$ (6.12)

By arguments based upon the countability of $(L_t : t \geq 1)$, the compacity of the spaces of probability measures defined on them, and a diagonal procedure, it can be shown that there exist subsequences $t' \to \infty$ such that the following limits exist:

$$\lim_{t' \to \infty} \theta_{t,t'}\nu_{t'} = \tau_t \text{ for all } t \geq 1. \qquad (6.13)$$

By classical results, for each one of these subsequences $t' \to \infty$ there exists a unique measure τ defined on Ω_∞ such that:

$$\theta_{t,\infty}\tau = \tau_t \text{ for all } t \geq 1 \qquad (6.14)$$

Every one of these measures τ is called a thermodynamic limit, as $L_t \to L_\infty$, of the Gibbs ensembles $(\nu_t : t \geq 1)$. It turns out that every one of the above thermodynamic limits τ is a Gibbs state for the Hamiltonian H. Refer to sections 1.4, 1.5, 1.6 of reference [Ru] where the above discussion is carried out in detail.

Recall that any thermodynamic limit τ is determined by the values it takes on the finite subsets Ω_t (this happens for any probability measure defined in Ω_∞). More precisely τ is determined by the values:

$$\tau\{\sigma \in \Omega_\infty : \sigma_{L_s} = \tilde{\sigma}_s\}, \text{ for } \tilde{\sigma}_s \in \Omega_s, \ s \geq 1$$

From the definitions we have:

$$\tau\{\sigma \in \Omega_\infty : \sigma_{L_s} = \tilde{\sigma}_s\} = \tau_s\{\tilde{\sigma}_s\} = \lim_{t' \to \infty} \nu_{t'}\{\sigma \in \Omega_{t'} : \sigma_{L_s} = \tilde{\sigma}_s\} \qquad (6.15)$$

where $t' \to \infty$ is a subsequence satisfying (6.13).

Then to describe the probability measure we are led to study the properties of the set of limit points of the sequence $(\nu_t\{\sigma \in \Omega_t : \sigma_{L_s} = \tilde{\sigma}_s\} : t \geq 1)$. In the next section we shall find the equations which the limit points of the above sequence satisfy. From expression (6.15) we deduce that the values $\tau\{\sigma \in \Omega_\infty : \sigma_{L_s} = \tilde{\sigma}_s\}$ will also satisfy these equations. For the case $q = 2$ i.e. $L_\infty = \mathbb{Z}$ we shall prove that the set of limit points of $(\nu_t\{\sigma \in \Omega_t : \sigma_{L_s} = \tilde{\sigma}_s\} : t \geq 1)$ is a singleton; then we shall conclude that the set of thermodinamic limits $\{\tau\}$ for $L_t \to L_\infty$ is also a singleton.

6.5. Evolution Equations

Let ν_t be the Gibbs ensemble on Ω_t. The one-site Gibbs state on the node $l = 0$ is given by:

$$\xi_t^0(i) = \nu_t\{\sigma \in \Omega_t : \sigma(0) = i\} \text{ for } i \in I \tag{6.16}$$

From formula (6.8) it can be shown that it satisfies the equality:

$$\xi_t^0(i) = Z_t^{-1}(g_t(i))^q \exp\left([N, i]\right) \tag{6.17}$$

We shall evaluate ν_t on the configurations $\tilde{\sigma}_s \in \Omega_s$ for $2 \le s \le t$ fixed. We have:

$$\xi_t^s(\tilde{\sigma}_s) = \nu_t\{\sigma \in \Omega_t : \sigma_{L_s} = \tilde{\sigma}_s\}. \tag{6.18}$$

By definition:

$$\xi_t^s(\tilde{\sigma}_s) = Z_t^{-1} \sum_{\{\sigma \in \Omega_t : \sigma_{L_s} = \tilde{\sigma}_s\}} \exp\left(-H_t(\sigma)\right)$$

and it is not hard to prove that:

$$\xi_t^s(\tilde{\sigma}_s) = Z_t^{-1} \exp\left(-H_s(\tilde{\sigma}_s)\right) \prod_{(l_1,\ldots,l_s) \in L_s \setminus L_{s-1}} (g_{t-s}(\tilde{\sigma}_s(l_1,\ldots,l_s)))^{q-1}$$

Apply the same formula to $(\tilde{\sigma}_s)_{L_{s-1}} \in \Omega_{s-1}$, the restriction of $\tilde{\sigma}_s$ to L_{s-1}, to get:

$$\xi_t^{s-1}((\tilde{\sigma}_s)_{L_{s-1}}) =$$
$$Z_t^{-1} \exp\left(-H_{s-1}((\tilde{\sigma}_s)_{L_{s-1}})\right) \prod_{(l_1,\ldots,l_{s-1}) \in L_{s-1} \setminus L_{s-2}} (g_{t-s+1}(\tilde{\sigma}_s(l_1,\ldots,l_{s-1})))^{q-1}$$
$$\tag{6.19}$$

From formula (6.9) we obtain:

$$\xi_t^{s-1}(\tilde{\sigma}_{s-1}) = Z_t^{-1} \exp\left(-H_{s-1}((\tilde{\sigma}_s)_{L_{s-1}})\right).$$

$$\prod_{(l_1,\ldots,l_s) \in L_s \setminus L_{s-1}} \left(\sum_{j \in I} g_{t-s}(j)^{q-1} \exp\left([N + K\tilde{\sigma}_s(l_1,\ldots,l_{s-1}), j]\right)\right)$$

Then the μ_t-conditional probability that $\tilde{\sigma}_s$ happens with respect to the condition that $(\tilde{\sigma}_s)_{L_{s-1}}$ has occurred is:

$$\nu_t\{\sigma \in \Omega_t : \sigma_{L_s} = \tilde{\sigma}_s | \sigma_{L_{s-1}} = (\tilde{\sigma}_s)_{L_{s-1}}\} =$$
$$\prod_{(l_1,\ldots,l_s) \in L_s \setminus L_{s-1}} \left(\frac{g_{t-s}(\tilde{\sigma}_s(l_1,\ldots,l_s))^{q-1} \exp([N + K\tilde{\sigma}_s(l_1,\ldots,l_{s-1}), \tilde{\sigma}_s(l_1,\ldots,l_s)])}{\sum_{j \in I}(g_{t-s}(j))^{q-1} \exp([N + K\tilde{\sigma}_s(l_1,\ldots,l_{s-1}), j])}\right)$$
$$\tag{6.20}$$

In the case $s = 1$, $l = (l_1, ..., l_{s-1}) = 0$ is the central node and then the above equation reads:

$$\nu_t\{\sigma \in \Omega_t : \sigma_{L_1} = \tilde{\sigma}_1 | \sigma(0) = \tilde{\sigma}_1(0)\} =$$

$$\prod_{l_1 \in Q_1} \left\{ \frac{g_{t-1}(\tilde{\sigma}_1(l_1))^{q-1} \exp\left([N + K\tilde{\sigma}_1(0), \tilde{\sigma}_1(l_1)]\right)}{\sum_{j \in I} g_{t-1}(j)^{q-1} \exp\left([N + K\tilde{\sigma}_1(0), j]\right)} \right\}$$

6.6. The One-Site Distribution of the Thermodynamic Limits

We shall obtain the period-2 equation that is satisfied by the one-site distribution of thermodynamic limits on the Bethe lattice (they were first set forth in [GM4]). To this purpose, we write:

$$x_i(t) = (\xi_t^0(i))^{\frac{1}{q}} = (\nu_t\{\sigma \in \Omega_t : \sigma(0) = i\})^{\frac{1}{q}} \text{ for } i \in I \qquad (6.21)$$

From (6.17) we have:

$$x_i(t) = Z_t^{-\frac{1}{q}} g_t(i) \exp\left(\frac{1}{q}[N, i]\right)$$

Now by using recursive equation (6.9) we get:

$$x_i(t) = (Z_t^{-1} Z_{t-1}^{q-1})^{\frac{1}{q}} (\sum_{j \in I} \exp\left(K[i, j] + [N, i]\right) \exp\left(\frac{1}{q}[N, i]\right) \cdot$$

$$(x_j(t-1))^{q-1} \exp\left(-\frac{q-1}{q}[N, j]\right))$$

Hence we obtain:

$$x_i(t) = (Z_k^{-1} Z_{t-1}^{q-1})^{\frac{1}{q}} \sum_{j \in I} (\exp\left(K[i+j] + \frac{1}{q}[N, i+j]\right))(x_j(t-1))^{q-1} \qquad (6.22)$$

Denote $x(t) = (x_i(t) : i \in I) \in \mathbb{R}_+^{|I|}$. For any vector $x \in \mathbb{R}_+^{|I|}$ define its p-norm and its ρ-power by:

$$\|x\|_p = (\sum_{i \in I} x_i^p)^{\frac{1}{p}} \text{ for } p \geq 1, \quad \varphi_\rho(x) = x^\rho = (x_i^\rho : i \in I) \text{ for } \rho \in \mathbb{R} \qquad (6.23)$$

(Refer to example 5.4 in section 5.8).

Now define the matrix $A = (a_{ij} : i, j \in I)$ by:

$$a_{ij} = \exp\left(K[i,j] + \frac{1}{q}[N, i+j]\right) \qquad (6.24)$$

Since $A > 0$, it assigns to any non-negative vector different from zero $x \in \mathbb{R}_+^{|I|} \setminus \{0\}$ an element $Ax \in \mathbb{R}_+^{|I|} \setminus \{0\}$: $(Ax)_i = \sum_{j \in I} a_{ij} x_j$ for $i \in I$.

With the above notation we can write equation (6.22) as:

$$x(t) = (Z_t^{-1} Z_{t-1}^{q-1})^{\frac{1}{q}} (Ax(t-1)^{q-1})$$

Besides, the constraint $\sum_{i \in I} \xi_t^0(i) = 1$ for any $t \geq 1$ implies $\|x(t)\|_q = 1$. Then we get:

$$(Z_t^{-1} Z_{t-1}^{q-1})^{\frac{1}{q}} = \|Ax(t-1)^{q-1}\|_q^{-1}$$

Hence equation (6.23) can be put in the form:

$$x(t) = \frac{Ax(t-1)^{q-1}}{\|Ax(t-1)^{q-1}\|_q} \qquad (6.25)$$

Define the function:

$$h_p(x) = \frac{x}{\|x\|_p} \text{ if } x \neq 0, \ h(0) = 0 \text{ for } x \in \mathbb{R}^{|I|}, \ p \geq 1 \qquad (6.26)$$

Since $\varphi_p(x) = x^p$, we can write evolution (6.25) as:

$$x(t) = U(x(t-1)), \text{ with } U = h_q \circ A \circ \varphi_{q-1} \text{ being the transformation}$$
$$\text{acting on the unit positive set } C_q = \{x \in \mathbb{R}_+^{|I|} : \|x\|_q = 1\} \qquad (6.27)$$

Theorem 6.1. [GM4]. Let τ be a thermodynamic limit and $\tau_0 = (\tau_0(i) : i \in I)$ be the one-site distribution vector i.e. $\tau_0(i) = \tau\{\sigma \in \Omega_\infty : \sigma(0) = i\}$. Then the following period-2 equation holds:

$$\tau_0^{\frac{1}{q}} = U^2(\tau_0^{\frac{1}{q}}) \text{ where } U = h_q \circ A \circ \varphi_{q-1} \qquad (6.28)$$

with the matrix A and the functions φ_p, h_p as specified in (6.23), (6.24), (6.26).

Proof. The continuous functions A, φ_{q-1} map $\mathbb{R}_+^{|I|} \setminus \{0\}$ into itself and h_q maps continuously $\mathbb{R}_+^{|I|} \setminus \{0\}$ into $C_q = \{x \in \mathbb{R}_+^{|I|} : \|x\|_q = 1\}$. Hence $U : \mathbb{R}_+^{|I|} \setminus \{0\} \to C_q$ is also continuous.

From formula (6.15) we find that the vector τ_0 belongs to the set of limit points of the sequence $\{\xi_t^0 = x(t)^q : t \geq 1\}$. Let $t' = (t_r : r \geq 1)$ be a subsequence such that:

$$\tau_0 = \lim_{r \to \infty} x(t_r)^q$$

As C_q is a compact subset, the subsequence $(t_r - 1 : r \geq 2)$ contains a subsequence $t'' = (t'_s : s \geq 1)$ such that $\lim_{s \to \infty} x(t'_s)$ exists. Call $u(0) = \lim_{s \to \infty} x(t'_s)$ and denote $u(t) = U^t(u(0))$ for $t \geq 0$. From continuity we have $(u(1))^q = (U(u(0)))^q = \lim_{s \to \infty} (U(x(t'_s)))^q = \lim_{s \to \infty} x(t'_s)^q = \tau_0$.

On the closed convex subset $F = \mathbb{R}_+^{|I|}$ the function

$$f(x) = \varphi_{q-1} \circ h_q(x) = \left(\frac{x}{\|x\|_q}\right)^{q-1} \text{ if } x \neq 0, \ f(0) = 0 \tag{6.29}$$

is a subgradient of $g(x) = \|x\|_q$. Also, the matrix A is symmetric.

By the results of section 5.8, in particular by expression (5.58), the sequence $E(x(t)) = (-\|A\varphi_{q-1}(x(t))\|_q : t \geq 1)$ decreases with $t \geq 1$ to a finite quantity $M \leq 0$. Let $v(t) = A\varphi_{q-1}(u(t))$ for $t \geq 0$. By continuity $\|v(t)\| = -M$ for any $t \geq 0$.

Also by continuity of function $\Delta_t E = E(x(t)) - E(x(t-1)) = - < f(A\varphi(x(t-1))), A\varphi(x(t+1)) - A\varphi(x(t-1)) >$ we deduce $< f(v(2)), v(2) - v(0) >= 0$. Then:

$$\|v(2)\|_q = \|v(0)\|_q + < f(v(0)), v(2) - v(0) > \tag{6.30}$$

Now, from the equalities:

$$< f(v(0)), v(0) >= \|v(0)\|_q, \quad \|v(0)\|_q^{q-1} = \|v(0)^{q-1}\|_{q/(q-1)}$$

expression (6.30) becomes:

$$\|v(2)\|_q \|v(0)^{q-1}\|_{q/q-1} = < v(0)^{q-1}, v(2) >$$

This means the Holder inequality becomes an equality.

Then there exist $\lambda, \lambda' \neq 0$ such that $\lambda(v(2))^q = \lambda'(v(0)^{q-1})^{q/q-1}$ which implies $v(2) = \lambda'' v(0)$. Since $\|v(2)\|_q = \|v(0)\|_q$ we deduce $v(2) = v(0)$ i.e. $A\varphi_{q-1}(v(2)) = A\varphi_{q-1}(u(0))$. Apply h_q to the above equality to get $u(3) = u(1)$. Hence $(u(3))^q = (u(1))^q = \tau_0$, so we find $\tau_0^{\frac{1}{q}} = U^2(\tau_0^{\frac{1}{q}})$. ∎

For the special case $L_\infty = \mathbb{Z}$ we have:

Corollary 6.1. [Br]. If $q = 2$, i.e. $L_\infty = \mathbb{Z}$, the one-site distribution vector τ_0 of a thermodynamic limit is the unique probability vector which is a solution of the following Perron-Frobenius equation:

$$A\tau_0^{\frac{1}{2}} = \lambda \tau_0^{\frac{1}{2}} \text{ for } \lambda > 0 \tag{6.31}$$

The vector τ_0 turns out to be strictly positive.

Proof. From the last theorem, τ_0 satisfies the equation $\tau_0^{\frac{1}{2}} = U^2(\tau_0^{\frac{1}{2}})$. But φ_1 is the identity and $h_q \circ A \circ h_q \circ A = h_q \circ A^2$ on $\mathbb{R}_+^* \setminus \{0\}$, so $\lambda' \tau_0^{\frac{1}{2}} = A^2 \tau_0^{\frac{1}{2}}$, with $\lambda' = \|A^2 \tau_0^{\frac{1}{2}}\|_q$. Hence $x = \tau_0^{\frac{1}{2}}$ is a non-negative solution to the equation $A^2 = \lambda' x$ with $\lambda' > 0$, satisfying $\|x\|_2 = 1$. Since $A > 0$ we have $A^2 > 0$ then we can apply the Perron-Frobenius theory to deduce that $\tau_0^{\frac{1}{2}}$ is the unique non-negative vector of norm $\|x\|_2 = 1$ satisfying the above equation.

Now consider the equation $Ay = \lambda y$ with $\lambda > 0$. Perron-Frobenius theorem asserts that there always exists a non-negative solution of norm $\|y\|_2 = 1$ to this equation. Such solution will satisfy $A^2 y = \lambda^2 y$ with $\lambda > 0$. By unicity we deduce $y = \tau_0^{\frac{1}{2}}$, hence τ_0 satisfies (6.31). Also by Perron-Frobenius this solution $\tau_0^{\frac{1}{2}}$ is a strictly positive vector. ∎

With a large ferromagnetic interaction, i.e. $K >> 0$, and in the absence of a exterior magnetic field, i.e. $N = 0$, the one-site-distribution τ_0 satisfies a fixed point condition.

Corollary 6.2. [GM4]. Assume $N = 0$, $K \geq 0$. Define

$$c = \inf\{[i, i] : i \in I\}, \quad c' = \sup\{[i, j] : i \neq j \text{ in } I\}$$

Then if one of the following conditions is satisfied:

$$\begin{aligned} K(c - c') &> \log(|I| - 1) \quad \text{or} \\ K(c - c') &> \log(|I| - 1) \text{ and } \exists! i_0 \in I \text{ such that } c = [i_0, i_0] \end{aligned} \tag{6.32}$$

the one-site distribution vector τ_0 of a thermodynamic limit satisfies the fixed point equation:

$$\tau_0^{\frac{1}{q}} = U(\tau_0^{\frac{1}{q}}) \tag{6.33}$$

Proof. Assume $K(c - c') - \log(|I| - 1) > 0$. Define

$$\delta = (\exp (Kc)) - (|I| - 1) \exp (Kc') > 0$$

Then:

$$\gamma = \sum_{i,j \in I} a_{ij} z_i z_j \geq \sum_{i \in I} a_{ii} z_i^2 - \sum_{i \neq j} a_{ij} |z_i| |z_j|$$

$$\geq \delta \sum_{i \in I} z_i^2 + (\exp (Kc'))\{(|I| - 1) \sum_{i \in I} z_i^2 - \sum_{i \neq j} |z_i| |z_j|\}$$

$$\geq \delta \sum_{i \in I} z_i^2 + \frac{1}{2}(\exp (Kc')) \sum_{i \neq j} (|z_i| - |z_j|)^2$$

So $\gamma \geq 0$ and $\gamma = 0$ iff $z = 0$. Hence A is positive-definite and the result follows from Proposition 5.5.i. If the second condition of (6.32) is satisfied we put:

$$\delta' = \inf \{(\exp (K[i,i]))) - (\exp (Ki_0)) : i \neq i_0\}$$

which is strictly positive. Then:

$$\gamma \geq \delta' \sum_{i \in I \setminus \{i_0\}} z_i^2 + \frac{1}{2}(\exp (K\gamma')) \sum_{i \neq j} (|z_i| - |z_j|)^2 \geq 0.$$

Let $u = U(\tau_0^{\frac{1}{q}})$ and $z = \varphi_{q-1}(u) - \varphi_{q-1}(\tau_0^{\frac{1}{q}})$. By using the proof of Proposition 5.5.i we deduce $\gamma = \sum_{i,j \in I} a_{ij} z_i z_j \leq 0$. Then $\gamma = 0$, but this occurs iff $z_i = 0$ for every $r \in S \setminus \{r_0\}$. Since $\|u\|_q = \|\tau_0^{\frac{1}{q}}\|_q = 1$ we deduce that $u = \tau_0^{\frac{1}{q}}$. ∎

Remarks.

1.- If the spin set satisfies $I = \{-1,1\}$ or $I = \{0,1\}$, we deduce that fixed point condition (6.33) follows for any positive ferromagnetic interaction $K > 0$ and null exterior magnetic field $N = 0$. In fact, if $I = \{-1,1\}$ then $c = 1$, $c' = -1$ and the first condition of (6.32) is satisfied for any $K > 0$. When $I = \{0,1\}$ the second condition of (6.32) is satisfied by taking $i_0 = 0$.

2.- Suppose that all spin vectors arc unitary $[i,i] = 1$ for any $i \in I$. Then $c' = \inf \{\cos \theta(i,j) : i \neq j\}$ where $\theta(i,j)$ is the angle between spins i,j. The first condition in (6.32) reads $K > (\log |I| - 1)(1 - c')^{-1}$. Hence if $|I| = 2$ and the spin vectors are unitary and different the fixed point condition (6.33) is satisfied for any $K > 0$ and $N = 0$.

6.7. Distribution of the Thermodynamic Limits

Let $1 \leq s \leq t$. We shall study the ν_t-conditional probability that a state $\tilde{\sigma}_s \in \Omega_s = L_s^I$ happens in $\Omega_t = L_t^I$ given that $(\tilde{\sigma}_s)_{L_{s-1}}$ has occurred. This ν_t-conditional probability obeys equation (6.20) of section 6.5. We shall decompose this expression according to its factors. To purpose define:

$$\alpha_i(t|k) = \frac{(g_t(i))^{q-1} \exp\left([N+Kk,i]\right)}{\sum\limits_{j \in I} (g_t(j))^{q-1} \exp\left([N+Kk,j]\right)} \quad \text{for } k,i \in I,\ t \geq 1 \tag{6.34}$$

which corresponds to a one-step transition probability, measured with distribution ν_t, between two neighbouring sites.

Now equation (6.20) can be written in the form:

$$\nu_t\{\sigma \in \Omega_t : \sigma_{L_s} = \tilde{\sigma}_s |\sigma_{L_{s-1}} = (\tilde{\sigma}_s)_{L_{s-1}}\} =$$
$$\prod_{(l_1,\ldots,l_s) \in L_s \setminus L_{s-1}} \alpha_{\tilde{\sigma}_s(l_1,\ldots,l_s)}(t - s|\tilde{\sigma}_s(l_1,\ldots,l_{s-1})) \tag{6.35}$$

In order to describe the limit behaviour of the sequence:

$$\left(\nu_t\{\sigma \in \Omega_t : \sigma_{L_s} = \tilde{\sigma}_s |\sigma_{L_{s-1}} = (\tilde{\sigma}_s)_{L_{s-1}}\} : t \geq 1\right)$$

we shall study the behaviour of its factors $(\alpha_i(t|k) : t \geq 1)$.

Recall that $\sum\limits_{i \in I} \alpha_i(t|k) = 1$. Define the following functions:

$$v_t(k) = \sum_{j \in I} g_t(j)^{q-1} \exp\left([N+Kk,j]\right)$$

Definition (6.34) implies: $\alpha_i(t|k) = (v_t(k))^{-1}(g_t(k))^{q-1} \exp([N+Kk,i])$. Hence $g_{t-1}(j)^{q-1} = \alpha_j(t-1|k)v_{t-1}(k) \exp\left(-[N+Kk,j]\right)$.

By using recursive formula (6.9) we find the equation:

$$(\alpha_i(t|k)v_t(k) \exp\left(-[N+Kk,i]\right))^{\frac{1}{q-1}} = \sum_{j \in I} \alpha_j(t-1|k)v_{t-1}(k) \exp\left([K(i-k),j]\right)$$

Take the following vector $y(t|k) = (y_i(t|k) : i \in I)$:

$$y_i(t|k) = (\alpha_i(t|k) \exp\left(-[N+qKk,i]\right))^{\frac{1}{q-1}} \tag{6.36}$$

It satisfied the relation:

$$y_i(t|k) = \gamma_t(k) \sum_{j \in I} (y_j(t-1|k)^{q-1} \{ \exp K([i,j] - [k,i+j]) \} \{ \exp ([N+qKk,j]) \}$$

(6.37)

where $\gamma_t(k) = v_t(k)^{-\frac{1}{q-1}} v_{t-1}(k)$.

Consider the vector $\mu_{(k)} = (\mu_{i,(k)} : i \in I)$, and the matrices $B_{(k)} = (b_{ij,(k)} : i,j \in I)$, $C_{(k)} = (c_{ij,(k)} : i,j \in I)$, indexed by $k \in I$ and defined as follows:

$$\mu_{i,(k)} = \exp([N+qKk,i])$$

(6.37)

$$b_{ij,(k)} = \exp(K([i,j]-[k,i+j]))$$

(6.39)

$$c_{ij,(k)} = b_{ij,(k)} \mu_{j,(k)}$$

(6.40)

Equality (6.37) implies the following recursive relation:

$$y_i(t|k) = \gamma_t(k) \sum_{j \in I} c_{ij,(k)} y_j(t-1|k)^{q-1}$$

or equivalentely written in vector terms:

$$y(t|k) = \gamma_t(k) C_{(k)} y(t-1|k)^{q-1}$$

(6.41)

From the equality $\sum_{i \in I} \alpha_i(t|k) = 1$ and definitions (6.36), (6.38) we get:

$$\sum_{i \in S} \mu_{i,(k)} y_i(t|k)^{q-1} = 1$$

(6.42)

On the set of real functions defined on I consider the following p-norm ($p \geq 1$), which uses the weights given by the vector $\mu_{(k)}$:

$$|||y|||_{p,(k)} = \left(\sum_{i \in S} \mu_{i,(k)} y_i^p \right)^{\frac{1}{p}} \text{ for } p \geq 1, \ k \text{ fixed in } I$$

(6.43)

With this notation equation (6.42) can be written as:

$$|||y(t|k)|||_{q-1,(k)} = 1$$

(6.44)

Apply this equality in (6.41) to get:

$$y(t|k) = \frac{C_{(k)} y(t-1|k)^{q-1}}{|||C_{(k)} y(t-1|k)^{q-1}|||_{q-1,(k)}}$$

(6.45)

Let us study an evolution which is similar to the previous one:

$$x(t) = \frac{C_{(k)} x(t-1)^{q-1}}{|||C_{(k)} x(t-1)^{q-1}|||_{q,(k)}} \tag{6.46}$$

It can also be written as:

$$x(t) = \tilde{U}_{(k)} x(t-1) \text{ with } \tilde{U}_{(k)} = \bar{h}_{q,(k)} \circ C_{(k)} \circ \varphi_{q-1} \tag{6.47}$$

where $\varphi_\rho(x) = x^\rho$ (see (6.23)) and $\bar{h}_{p,(k)}$ is the following function:

$$\bar{h}_{p,(k)}(x) = \frac{x}{|||x|||_{p,(k)}} \text{ if } x \neq 0, \ \bar{h}_{p,(k)}(0) = 0 \tag{6.48}$$

Now $f = \varphi_{q-1} \circ \bar{h}_{q,(k)}$ satisfies $f(x) = \frac{x^{q-1}}{|||x|||_{q,(k)}^{q-1}}$ if $x \neq 0$, $f(0) = 0$. Then f is a subgradient of the norm $|||x|||_{q,(k)}$ with respect to the inner product $< \ , \ >_{(k)}$ i.e.:

$$|||x|||_{q,(k)} \geq |||z|||_{q,(k)} + < f(z), x - z >_{(k)}$$
$$\text{where } < z, z' >_{(k)} = \sum_{i \in I} z_i z_i' \mu_{i,(k)} \tag{6.49}$$

Since $|||x|||_{q,(k)}$ is a norm we also have: $< x, f(x) >_{(k)} = |||x|||_{q,(k)}$.
The matrix $B_{(k)}$ given by (6.39) is symmetric, then:

$$\mu_{i,(k)} c_{ij,(k)} = \mu_{j,(k)} c_{ji,(k)} = \mu_{i,(k)} \mu_{j,(k)} b_{ij,(k)}$$

which is equivalent to saying that the matrix $C_{(k)}$ is $<>_{(k)}$-self adjoint i.e.:

$$< C_{(k)} x, z >_{(k)} = < x, C_{(k)} z >_{(k)}$$

Hence from the discussion in sections 5.8 and 5.9, and by an entirely analogous proof to that given for Theorem 6.1 we get:

Lemma 6.1. [M2]. Any limit point x^* of the sequence $(x(t) : t \geq 1)$ whose evolution is given by equation (6.46) satisfies the period-2 equation:

$$x^* = \tilde{U}_{(k)}^2 x^* \quad \blacksquare$$

Now let us show:

Theorem 6.2. [M2]. Any limit point $y_{(k)}^* = (y_{i,(k)}^* : i \in I)$ of the sequence $(y(t|k) : t \geq 1)$ whose evolution is given by equation (6.45) satisfies the period-2 equation:

$$y_{(k)}^* = (U_{(k)})^2 y_{(k)}^* \tag{6.50}$$

where the transformation $U_{(k)}$ depending on $k \in I$ is given by:

$$U_{(k)} = \bar{h}_{q-1,(k)} \circ C_{(k)} \circ \varphi_{q-1} \tag{6.51}$$

Proof. Let $(y(t|k) : t \geq 1)$ be the sequence given by evolution (6.45). Now take $\tilde{y}(t|k) = \frac{y(t|k)}{|||y(t|k)|||_{q,(k)}}$ for $t \geq 1$. It is easy to show that $(\tilde{y}(t|k) : t \geq 1)$ satisfies evolution (6.46). Take a limit point $y_{(k)}^*$ of $(y(t|k) : n \geq 1)$, then $\tilde{y}_{(k)}^* = \frac{y_{(k)}^*}{|||y_{(k)}^*|||_{q,(k)}}$ is a limit point of $(\tilde{y}(t|k) : t \geq 1)$. According to Lemma 6.1 the point $\tilde{y}_{(k)}^*$ satisfies the equation $\tilde{y}_{(k)}^* = \tilde{U}_{(k)}^2 \tilde{y}_{(k)}^*$. Then we deduce that $y_{(k)}^* = \bar{h}_{q-1,(k)}(\tilde{U}_{(k)}^2 y_{(k)}^*)$. From the equality $\bar{h}_{q-1,(k)} \circ \tilde{U}_{(k)}^2 = U_{(k)}^2$ we conclude (6.50). ∎

Now let us prove that the marginal distributions $\tau\{\sigma \in \Omega_\infty : \sigma_{L_s} = \tilde{\sigma}_s\}$ of the thermodynamic limit satisfies a period-2 equation for $\tilde{\sigma}_s \in \Omega_s$ fixed. Recall:

$$\tau_s\{\tilde{\sigma}_s\} = \tau\{\sigma \in \Omega_\infty : \sigma_{L_s} = \tilde{\sigma}_s\}.$$

According to (6.15) we have: $\tau_s\{\tilde{\sigma}_s\} = \lim_{t' \to \infty} \nu_{t'}\{\sigma \in \Omega_{t'} : \sigma_{L_s} = \tilde{\sigma}_s\}$ for some subsequence $t' \to \infty$. By developping $\nu_{t'}\{\sigma \in \Omega_{t'} : \sigma_{L_s} = \tilde{\sigma}_s\}$ we obtain:

$$\nu_{t'}\{\sigma \in \Omega_t : \sigma_{L_s} = \tilde{\sigma}_s\} = \prod_{r=1}^{s} \nu_{t'}\{\sigma \in \Omega_{t'} : \sigma_{L_r} = (\tilde{\sigma}_s)_{L_r}, |\sigma_{L_{r-1}} = (\tilde{\sigma}_s)_{L_{r-1}}\}$$
$$\cdot \nu_{t'}\{\sigma \in \Omega_{t'} : \sigma(0) = \tilde{\sigma}_s(0)\}$$

Then from equation (6.35):

$$\nu_{t'}\{\sigma \in \Omega_{t'} : \sigma_{L_s} = \tilde{\sigma}_s\} = \prod_{r=1}^{s} \prod_{(l_1,\ldots,l_r) \in L_r \backslash L_{r-1}} \alpha_{\tilde{\sigma}_s(l_1,\ldots,l_r)}(t' - r|\tilde{\sigma}_s(l_1,\ldots,l_{r-1}))$$
$$\cdot \nu_{t'}\{\sigma \in \Omega_{t'} : \sigma(0) = \tilde{\sigma}_s(0)\} \tag{6.52}$$

For $r = 0$ we define: $\alpha_{\tilde{\sigma}_s(l_1,\ldots,l_r)}(t' - r|\tilde{\sigma}_s(l_1,\ldots,l_{r-1}))$
$$= \nu_{t'}\{\sigma \in \Omega_{t'} : \sigma(0) = \tilde{\sigma}_s(0)\}$$

(recall that the node $(l_1, ..., l_r) = 0$ for $r = 0$ and the expression $\tilde{\sigma}_s(l_1, ..., l_{r-1})$ correspond to a void restriction for $r = 0$).

Then (6.52) can be written as:

$$\nu_{t'}\{\sigma \in \Omega_{t'} : \sigma_{L_s} = \tilde{\sigma}_s\} = \prod_{r=0}^{s} \prod_{(l_1,...,l_r) \in L_r \setminus L_{r-1}} \alpha_{\tilde{\sigma}_s(l_1,...,l_r)}(t' - r|\tilde{\sigma}_s(l_1, ..., l_{r-1}))$$

where $L_{-1} = \phi$ is the empty set. Hence:

$$\tau_s\{\tilde{\sigma}_s\} = \lim_{t' \to \infty} \left(\prod_{r=0}^{s} \prod_{(l_1,...,l_r) \in L_r \setminus L_{r-1}} \alpha_{\tilde{\sigma}_s(l_1,...,l_r)}(t' - r|\tilde{\sigma}_s(l_1, ..., l_{r-1})) \right)$$

for any $s \geq 0$

Since this limit exists for any $\tilde{\sigma}_s \in \Omega_s$ and any $s \geq 0$, all the following limits exist as well:

$$\alpha_i^*(r; k) = \lim_{t' \to \infty} \alpha_i(t' - r|k) \text{ for any } i, k \in I \text{ and } r \geq 0 \qquad (6.53)$$

Recall $\tau_0 = (\alpha_i^*(0; k) : i \in I)$ is the one site distribution which does not depend on the state $k \in I$. Hence:

$$\tau_s\{\tilde{\sigma}_s\} = \prod_{r=0}^{s} \prod_{(l_1,...,l_r) \in L_r \setminus L_{r-1}} \alpha_{\tilde{\sigma}_s(l_1,...,l_r)}^*(r; \tilde{\sigma}_s(l_1, ..., l_{r-1})) \qquad (6.54)$$

A thermodynamic limit τ is biunivocally determined by the set of marginal distributions $(\tau_s\{\tilde{\sigma}_s\} : \tilde{\sigma}_s \in \Omega_s, s \geq 0)$, and this sequence is in turn in a one-to-one correspondence with the following sequence of vectors in $\mathbb{R}^{|I|}$:

$$\tau^*(y^*(r; k) : k \in I, r \geq 0) \text{ with } y^*(r; k) = (\alpha_i^*(r; k) : i \in I) \qquad (6.55)$$

The last class of vectors are limit points of evolutions of the type (6.25) for $r = 0$ and (6.45) for $r \geq 1$. Then they satisfy the following period-2 equation:

$$y^*(r; k) = U_{(r;k)}^2 y^*(r; k)$$
$$\text{where } U_{(0,k)} = U, \quad U_{(r;k)} = U_{(k)} \text{ if } r \geq 1 \qquad (6.56)$$

(U and $U_{(k)}$ are the transformation given in expressions (6.27) and (6.51) respectively).

For the representation τ^* given in (6.55) define the transformation:

$$\hat{U}(\tau^*) = (U_{(r;k)} y^*(r;k) : k \in I, r \geq 0) \tag{6.57}$$

From equality (6.56) and the definitions we can establish:

Theorem 6.3. [M2]. A thermodynamic limit τ satisfies the following period-2 equation:

$$\tau^* = \hat{U}^2(\tau^*) \tag{6.58}$$

where τ^* is given in (6.55) and $\hat{U}$ in (6.57). ■

For the special case $L_\infty = \mathbb{Z}$ we prove the unicity of the thermodynamic limit τ by showing that the vectors of its representation satisfy some Perron-Frobenius equations.

Corollary 6.3. [Br,M2]. If $q = 2$ i.e. $L_\infty = \mathbb{Z}$ the thermodynamic limit τ is unique. Furthermore if $\tau^* = (y^*(r;k) : k \in I, r \geq 0)$ is its representation, then each one of the vectors $y^*(r;k)$ satisfies the Perron-Frobenius equation:

$$C_{(k)} y^*(r;k) = \lambda y^*(r;k) \text{ with } \lambda > 0, y^*(r;k) \geq 0, \||y^*(r;k)\||_{q,(k)} = 1$$
$$\text{for } r \geq 1 \tag{6.59}$$
$$A y^*(0;k) = \lambda y^*(0;k) \text{ with } \lambda > 0, y^*(0;k) \geq 0, \|y^*(0;k)\|_1 = 1$$

The vectors $y^*(r;k)$ are strictly positive $\forall k \in I$, $r \geq 0$. Then τ gives every finite configuration strictly positive weight, i.e. $\tau\{\sigma \in \Omega_\infty : \sigma_{L_s} = \tilde{\sigma}_s\} > 0$ for any $\tilde{\sigma}_s \in \Omega_s$, $s \geq 0$.

Proof. The second equation of (6.59) is the same as equation (6.31), because $\tau_0 = y^*(0;h)$. On the other hand, from Theorem 6.2 we get that in the case $q = 2$ each vector $y^*(r;k)$ for $r \geq 1$, $k \in I$ satisfies the equation: $\lambda' y^*(r;k) = C_{(k)}^2 y^*(r;k)$ with $\lambda' > 0$. Then by the same discussion as in the proof of Corollary 6.1 we deduce (6.59). From $C_{(k)} > 0$ and Perron-Frobenius theorem we deduce that the vectors are strictly positive. Finally expression (6.54) implies that τ gives strictly positive weight to every finite configuration. ■

6.8. Period ≤ 2 Limit Orbits of Some Non Linear Dynamics on $I\!R_+^s$

Some other results concerning the period-2 limit orbits for non-linear dynamics on $I\!R_+^{|I|}$ can be established following the proof of Theorem 6.1.

Let I be finite. We assume the coefficients of the matrix $A = (a_{ij} : i, j \in I)$ to be non-negative, which we denote $A \geq 0$. We also suppose that A is chain-symmetric (see Definition 5.3). In Lemma 6.2 we search for conditions weaker than the strong condition $A > 0$ in order that A maps $I\!R_+^{|I|} \setminus \{0\}$ into itself.

According to Proposition 5.9, A is chain-symmetric if there exists a positive vector $\mu = (\mu_i : i \in I) > 0$ such that:

$$A \text{ is } \mu\text{-self-adjoint i.e. } < Ax, y >_\mu = < x, Ay >_\mu$$

$$\text{where } < x, y >_\mu = \sum_{i \in I} \mu_i x_i y_i$$

In the case where the matrix A is also irreducible this positive vector μ is unique, up to a positive constant. In the irreducible case the matrix A has a well defined period, that is, the period $\eta_i = \text{u.c.d. } \{n : a_{ii}^{(n)} > 0\}$ is common to all the states $i \in I$ (recall that $a_{ii}^{(n)}$ is the (i,i)-term of matrix A^n).

Lemma 6.2. Let $A \geq 0$ be irreducible and chain symmetric. Then $Ax \neq 0$ if $x \in I\!R_+^{|I|} \setminus \{0\}$ and A is a matrix of period 1 or 2.

Proof. Let $x \in I\!R_+^{|I|} \setminus \{0\}$ then there exists $i_0 \in I$ such that $x_{i_0} \neq 0$. If $Ax = 0$ then $a_{ji_0} = 0$ for any $j \in I$. Let μ be a strictly positive vector such that A is $<>_\mu$-self adjoint, then $\mu_i a_{ij} = \mu_j a_{ji}$ for any pair $i, j \in I$. Then we deduce that $a_{i_0 j} = 0$ for any $j \in I$. Hence the row and column i_0 are null, which contradicts the irreducibility of A.

To prove that A is of period 1 or 2 it suffices to show that $a_{ii}^{(2n)} > 0$ for any $n > 0$, $i \in I$. By putting $a_{jj}^{(0)} = 1$ for any $j \in I$ we have:

$$a_{ii}^{(2n)} \geq \sum_{j \in S} a_{ij} a_{jj}^{(2n-2)} a_{ij} \text{ for any } n \geq 1, \ i \in I$$

Let n be the smallest integer such that for some $i \in I$, $a_{ii}^{(2n)} = 0$. Since $a_{jj}^{(2n-2)} > 0$ for any $j \in I$ we get $a_{ij} a_{ji} = 0 \ \forall j \in I$. The $< >_\mu$-self adjointness property of A implies $a_{ij} = 0 = a_{ji}$ for any $j \in I$, which again contradicts the irreducibility of A. ■

Lemma 6.3. Let $A \geq 0$ be chain symmetric such that $A(\mathbb{R}_+^{|I|} \setminus \{0\}) \subset \mathbb{R}_+^{|I|} \setminus \{0\}$ (for instance if it is irreducible). Take the transformation:

$$Ux = \frac{Ax^p}{\|Ax^p\|} \quad \text{for } x \in \mathbb{R}_+^{|I|} \setminus \{0\}, \ p > 0, \ U(0) = 0. \tag{6.60}$$

where $\| \ \|$ is some norm in $\mathbb{R}^{|I|}$.

Then any limit point x^* of this evolution (i.e. $x^* = \lim_{s \to \infty} U^{t_s} x_0$ for some $x_0 \in \mathbb{R}_+^{|I|} \setminus \{0\}$ and some subsequence $(t_s) \underset{s \to \infty}{\longrightarrow} \infty$) satisfies:

$$x^* = U^2 x^*$$

Proof. We can write $U = h \circ A \circ \varphi_p$ where $\varphi_p(x) = x^p$ for $x \geq 0$ and $h(x) = \frac{x}{\|x\|}$ if $x \neq 0$, $h(0) = 0$. Let us first consider the case $\| \ \| = \| \ \|_{p+1}$. The proof of Theorem 6.1 implies the above lemma in this case, i.e. when h is h_{p+1} (given in (6.26)) and U is $\tilde{U} = h_{p+1} \circ A \circ \varphi_p$.

The result extends to any norm $\| \ \|$ by taking into account the following equalities on the set $\mathbb{R}_+^{|I|} \setminus \{0\}$: $h_{p+1} \circ h = h_{p+1}$, $h \circ h_{p+1} = h$, $U \circ h_{p+1} = \tilde{U} \circ h = \tilde{U}$, $h_{p+1} \circ U = h_{p+1} \circ \tilde{U}$, $h \circ U = h \circ \tilde{U}$. In fact induction gives $h_{p+1} \circ U^n = \tilde{U}^n$, $h \circ \tilde{U}^n = U^n$ on $\mathbb{R}_+^{|I|} \setminus \{0\}$. Then if x^* is a limit point for transformation U acting on $\mathbb{R}_+^{|I|} \setminus \{0\}$ the point $h_{p+1}(x^*)$ will be a limit point for transformation $\tilde{U}$ acting on $\mathbb{R}_+^{|I|} \setminus \{0\}$. Hence $h_{p+1}(x^*) = \tilde{U}^2 h_{p+1}(x^*)$. By the above equalities we conclude $x^* = U^2 x^*$. ∎

Let us summarize the results obtained for the nonlinear evolution of probability vectors.

Corollary 6.3. Let $A \geq 0$ be chain symmetric such that $A(\mathbb{R}_+^{|I|} \setminus \{0\}) \subset \mathbb{R}_+^{|I|} \{0\}$. Denote $C_1 = \{x \in \mathbb{R}^{|I|} : x \geq 0, \|x\|_1 = 1\}$ the simplex of probability vectors. Let $p > 0$, then the following evolutions on C_1:

$$Ux = \frac{Ax^p}{\|Ax^p\|_1} \tag{6.61}$$

$$\bar{U}x = \frac{(Ax^{p/p+1})^{p+1}}{\|Ax^{p/p+1}\|_{p+1}^{p+1}} \tag{6.62}$$

have only limit probability vectors whose orbits are of period 1 or 2.

Proof. For the transformation U of (6.61) the result follows from the last lemma. Now consider $\bar{U}$ given by (6.62). From the equality $\|(Ax^{p/p+1})^{p+1}\|_1 = \|Ax^{p/p+1}\|_{p+1}^{p+1}$ we conclude $\|\bar{U}x\|_1 = 1$, so $\bar{U}$ acts on C_1. Now $\bar{U}x = \bar{h} \circ A \circ \varphi_{p/p+1}$ where $\bar{h}(x) = \frac{x^{p+1}}{\|x\|_{p+1}^{p+1}}$. As $\bar{f}(x) = \varphi_{p/p+1} \circ \bar{h}(x) = \varphi_p \circ h_{p+1}(x)$ is a subgradient of $\| \ \|_{p+1}$ we use the same proof as in Theorem 6.1 to conclude the result. ∎

Recall that the transformation U of (6.61) is a direct generalization of a Markov chain because it corresponds to the evolution:

$$x_i(t+1) = (Ux(t))_i = \frac{\sum\limits_{j \in I} a_{ij} x_j^p(t)}{\sum\limits_{i \in I} \sum\limits_{j \in I} a_{ij} x_j^p(t)}$$

Hence it is convenient to write the above result in the following form.

Corollary 6.4. Let $A \geq 0$ be irreducible and chain symmetric . Let $p > 0$ then any limit probability vector x^* of the above transformation U has period 1 or 2, i.e. it satisfies the equation $x^* = U^2 x^*$.

Proof. By Lemma 6.2. ∎

Now we shall study other non linear evolutions of probability vectors that also give rise to limit points of period ≤ 2. We study functions that generalize the family $\varphi(x) = x^p$ and we normalize their actions with the Orlicz norm that they induce.

Let $\phi : \mathbb{R}_+ \to \mathbb{R}_+$ be a non decreasing right continuous function satisfying the following conditions:

$$\phi(0) = 0, \ \phi(u) > 0 \text{ when } u > 0 \text{ and } \phi(\infty) = \lim_{u \to \infty} \phi(u) = \infty \qquad (6.63)$$

For such a function define $G_\phi(u) = \int_0^u \phi(v)dv$.

For I finite G_ϕ defines a norm on $\mathbb{R}^{|I|}$ called the Orlicz norm (see reference [KR] for a proof that it is a norm):

$$\|x\|_\phi = \inf\{u > 0 : \sum_{i \in I} G_\phi(\frac{|x_i|}{u}) \leq 1\}$$

If $x \neq 0$ the norm $\|x\|_\phi$ is the unique $u > 0$ satisfying the equality:

$$\sum_{i \in I} G_\phi \left(\frac{|x_i|}{u} \right) = 1, \quad u = \|x\|_\phi \tag{6.64}$$

Recall that the class of functions $\phi_p(u) = \frac{1}{p+1} u^p$, $p > 0$, satisfy the above conditions. The Orlicz norm induced by ϕ_p is $\|x\|_{\phi_p} = \|x\|_{p+1}$.

On $\mathbb{R}_+^{|I|}$ we define:

$$\varphi_\phi(x) = (\phi(x_i) : i \in I), \quad h_\phi(x) = \frac{x}{\|x\|_\phi} \text{ if } x \neq 0, h_\phi(0) = 0 \tag{6.65}$$

Theorem 6.4. [GM4] Let $A \geq 0$ be irreducible and chain symmetric. Take a continuos strictly positive function ϕ satisfying the condition (6.63). Define the following transformation on $\mathbb{R}_+^{|I|}$:

$$U_\phi = h_\phi \circ A \circ \varphi_\phi \tag{6.66}$$

Then any limit point $x^* = \lim_{s \to \infty} U_\phi^{t_s} x(0)$ with $x(0) \in \mathbb{R}_+^{|I|}$, $t_s \xrightarrow[s \to \infty]{} \infty$ satisfies the period-2 equality:

$$x^* = U_\phi^2 x^*.$$

Proof. An easy computation on (6.64) shows that the norm $\| \ \|_\phi$ is differentiable on $\mathbb{R}_+^{|I|} \setminus \{0\}$ and its derivative $\hat{\varphi}_\phi(x) = (\frac{\partial}{\partial x_i} \|x\|_\phi : i \in I)$ is given by:

$$\hat{\varphi}_\phi(x) = \gamma \left(\frac{x}{\|x\|_\phi} \right) \varphi_\phi \left(\frac{x}{\|x\|_\phi} \right) \quad \text{where } \gamma(y) = \left(\sum_{i \in I} y_r \phi(y_r) \right)^{-1} > 0$$

Define $\hat{\varphi}_\phi(0) = 0$ so that the equality $h_\phi \circ A \circ \hat{\varphi}_\phi = h_\phi \circ A \circ \varphi_\phi$ holds on $\mathbb{R}_+^{|I|}$. Hence we can write $U_\phi = h_\phi \circ A \circ \hat{\varphi}_\phi$. Since $f = \hat{\varphi}_\phi \circ h_\phi = \hat{\varphi}_\phi$ on $\mathbb{R}_+^{|I|}$ the norm $\|x\|_\phi$ is a convex potential associated to f. From results of section 5.8, in particular from expression (5.59), we deduce that for any $x(0) \in \mathbb{R}_+^{|I|}$ the quantity $-\|A\hat{\varphi}_\phi(U_\phi^t x(0))\|_\phi$ decreases with $t \geq 1$ to a finite quantity $M \leq 0$.

Let $x^* = \lim_{s \to \infty} U_\phi^{t_s} x(0)$. By continuity there exists $z(0) \in \mathbb{R}_+^{|I|}$ such that $U_\phi z(0) = x^*$. Denote $z(t) = U_\phi^t z(0)$ for $t \geq 0$ and $v(t) = A\hat{\varphi}_\phi(z(t))$. So $\|v(t)\|_\phi = -M$ a constant ≥ 0 for any $t \geq 0$ and also $< \hat{\varphi}_\phi(v(2)), v(2) - v(0) >= 0$. We suppose $x(0) \neq 0$ (if not the result is evident because $x^* \neq 0$) then $-M > 0$.

Let $w(t) = v(t)/\|v(t)\|_\phi = -v(t)/M$. By definition of $\|\ \|_\phi$ we have $\sum\limits_{i \in I} G_\phi(w_i(t)) = 1$. From the equality

$$\hat\varphi_\phi(v(t)) = \hat\varphi_\phi(-Mw(t)) = \gamma(w(t))\varphi_\phi(w(t))$$

and the condition $\gamma(w(t)) \neq 0$ we deduce $< \varphi_\phi(w(2)),\ w(2) - w(0) >= 0$.

Now define $\tilde G_\phi(x) = \sum\limits_{i \in I} G_\phi(x_i)$, which is a strictly convex potential of φ_ϕ because ϕ is strictly increasing. From the above equalities we deduce:

$$\tilde G_\phi(w(2)) - \tilde G_\phi(w(0)) =< \varphi(w(2)), w(2) - w(0) >= 0$$

From the strict convexity of $\tilde G_\phi$ we get $w(2) = w(0)$, which implies $U_\phi^2 x^* = x^*$. ∎

A version of the above result as a nonlinear dynamics on probability vectors can be established. Let $G_\phi(x) = (G_\phi(x_i) : i \in S)$ and G_ϕ^{-1} be its inverse in $\mathbb{R}_+^{|I|}$. From the definition we have that G_ϕ maps biunivocally the set $C_\phi = \{x \in \mathbb{R}_+^{|I|} : \|x\|_\phi\}$ onto $C_1 = \{x \in \mathbb{R}_+^{|I|} : \|x\|_1 = 1\}$, this latter being the simplex of probability vectors.

Corollary 6.4. Let A, ϕ satisfy the hypothesis of Theorem 6.4. Then the limit probability vectors x^* of the evolution:

$$\tilde U_\phi(x) = G_\phi \circ h_\phi \circ A \circ \varphi_\phi \circ G_\phi^{-1}(x)$$

are of period 1 or 2, i.e. they satisfy $x^* = \tilde U_\phi^2 x^*$.

Proof. Note that $\tilde U_\phi = G_\phi \circ h_\phi \circ A \circ \tilde\varphi_\phi \circ G_\phi^{-1}$ and that it maps C_1 onto itself. Now take $h' = G_\phi \circ h_\phi$, $\varphi' = \hat\varphi_\phi \circ G_\phi^{-1}$; then $\varphi' \circ h' = \hat\varphi_\phi \circ h_\phi = \hat\varphi_\phi = \hat\varphi_\phi \circ h_\phi$ and by the proof of the last theorem we deduce the result. ∎

Remarks.

1. The result established in Lemma 6.3 also gives information about the dynamics of the transformation:

$$(\bar U_{(p)} x)_i = \sum\limits_{j \in I} a_{ij} x_j^p \quad \text{on } \mathbb{R}_+^{|I|}$$

In fact $\bar U_{(p)}(\lambda x) = \lambda^p \bar U_{(p)} x$ for $\lambda \geq 0$. Then the images of two points in the same ray also belong to the same ray. Hence $\bar U_{(p)}$ induces a transformation among the rays of $I\!R_+^I$ which is just the one given in (6.60) (we can use any norm $\| \ \|$). Then $\bar{U}_{(p)}$ accumulates only orbits of rays of period ≤ 2 in this action.

2. When $Ax = 0$ for some $x \in I\!R_+^{|I|} \setminus \{0\}$ most of the results obtained in section 6.8 -for instance for Lemma 6.3- are slighly modified. In fact if $U = h \circ A \circ \varphi$ is anyone of the transformations studied (for instance $\varphi = \varphi_\rho$, $h(x) = \frac{x}{\|x\|}$, with A chain-symmetric) then it can be shown that the orbits of the limit points are either of the form $(y, z, y, z, ...)$ or $(y, z, 0, 0, ...)$ for some $y, z \in I\!R_+^{|I|}$.

3. For non chain symmetric matrices our results are not necessarily true. For instance take:

$$A = \begin{pmatrix} 0 & 1 & 0 & \cdots & 0 \\ 0 & 0 & 1 & \cdots & 0 \\ \vdots & \vdots & \vdots & \vdots & \vdots \\ 0 & 0 & 0 & \cdots & 1 \\ 1 & 0 & 0 & \cdots & 0 \end{pmatrix}$$

is a cyclic matrix then we have the existence of finite orbits of length $|I|$ under anyone of the evolution $U = h \circ A \circ \varphi$ studied in section 6. These orbits of length $|I|$ continue to exist under strictly positive small perturbations of this matrix, i.e. for

$$A = \begin{pmatrix} \epsilon & 1 & \epsilon & \cdots & \epsilon \\ \epsilon & \epsilon & 1 & \cdots & \epsilon \\ \vdots & \vdots & \vdots & \vdots & \vdots \\ \epsilon & \epsilon & \epsilon & \cdots & \epsilon \\ 1 & \epsilon & \epsilon & \cdots & \epsilon \end{pmatrix}$$

where $0 < \epsilon << 1$.

References

[Bax] Baxter, R.J., *Exactly Solved Models in Statistical Mechanics*, Academic Press, 1982.

[Br] Brascamp, H.J., *Equilibrium States for a One Dimensional Lattice Gas*, Comm. Math. Phys, 21, 1971, 56-70.

[FV] Fannes, M., A. Verbeure, *On Solvable Models in Classical Lattice Systems*, Comm. Math. Phys. 96, 1984, 115-124.

[GM4] Goles, E., S. Martínez, *The One-Site Distributions of Gibbs States on Bethe Lattice are Probability Vectors of Period ≤ 2 for a Nonlinear Transformation*, J. Stat. Phys. 52, 1988, 267-285.

[KR] Krasnoselsky, M.A., Y.B. Rutitsky, *Convex Functions and Orlicz Spaces*, Industan Publ. Co., 1962.

[M1] Martínez, S., *Lyapunov Functionals on Bethe Lattice*, in Proceedings Workshop on Disordered Systems, Bogotá, World Sc. Publ. 1989, 22-37.

[M2] Martínez, S., *Cylinder Distribution of Thermodynamic Limit of Bethe Lattice*, in Proceedings Symposium on Non Equilibrium Structures, Valparaíso, D. Reidel Publ. Co., 1989.

[Ru] Ruelle, D., *Thermodynamic Formalism*, Addison-Wesley, 1978.

7. POTTS AUTOMATA

7.1. The Potts Model

The Potts model was introduced in Statistical Physics as a generalization of spin glasses [PoV,Pt,Wu]. Roughly described, the model comprises lattices where the spins may take several orientations rather than just two, as in the binary case (up and down). The Hamiltonian is:

$$H(x) = -\frac{1}{2} \sum_{(i,j) \in V} \delta_K(x_i, x_j); \quad x_j \in Q = \{0, ..., q-1\} \tag{7.1}$$

where Q is the finite set of orientations, $(i,j) \in V$ means the sites i and j are neighbours, and δ_K is the Kroeneker function: $\delta_K(u,v) = 1$ iff $u = v$. It is not difficult to see that (7.1) reduces to the Ising Hamiltonian (up to an additive constant) for $|Q| = 2$. In this chapter we suppose that the lattice interactions occur in a non-oriented graph $G = (I,V)$, where $I = \{1, ..., n\}$ is the set of sites, while the set of links V is assumed to be symmetric without loops: $(i,i) \notin V$.

For any $i \in I$ we define its neighbourhood as follows:

$$V_i = \{j \in I : (i,j) \in V\}$$

The fact that G is non-oriented is equivalent to saying that $(i,j) \in V$ iff $(j,i) \in V$, or that the incidence matrix of G is symmetric (i.e. $a_{ij} = 1$ iff $(i,j) \in V$). In this chapter we shall present a generalization of Hamiltonian (7.1) and we shall associate to it local rules that may cause the energy to (locally) decrease. We shall call them "compatible" rules. Then we shall associate a dynamics, either sequential or synchronous, to the model. In the sequential case we will prove that the dynamic behaviour is simple (it essentially leads to fixed points) but in the synchronous case we will evince its eventual complexity (i.e. there exist compatible rules with non trivial computing capabilities). Instead of the latter "negative" result we propose two classes of Compatible Networks which accept a Lyapunov functional imposed by the generalized Potts Hamiltonian. Hence they posses a simple dynamic behaviour: fixed points or two cycles. It is to be noted that in spite of the simplicity of compatible automata they are important in applications concerning image-smoothing, phase-unwrapping, and local strategies for some hard combinatorial optimization problems [G5,GMR,PoV].

7.2. Generalized Potts Hamiltonians and Compatible Rules

Let δ and δ^* be symmetric mappings of $I\!R \times I\!R$ into $I\!R$; i.e. $\delta(a, b) = \delta(b, a)$; $\delta^*(a, b) = \delta^*(b, a)$. The generalized Potts Hamiltonian is:

$$H(x) = -\frac{1}{2} \sum_{(i,j) \in V} \delta(x_i, x_j) + \sum_{i \in I} \delta^*(b_i, x_i) \qquad (7.2)$$

where $x_j \in Q$ and $b_i \in I\!R$.

Clearly (7.1) is the particular case where $\delta = \delta_K$ and $\delta^* = 0$.

The second term may be seen as an external magnetic field acting on the configurations supported by the graph G.

In order to have finite sums for infinity lattices we also suppose that there exists a quiescent state $q_0 \in Q$, such that $\delta(q_0, q_0) = \delta^*(b_i, q_0) = 0$. In this case, we can work with finite support configurations.

We may write H as follows:

$$H(x) = -\frac{1}{2} \sum_{i \in I} \sum_{j \in V_i} \delta(x_i, x_j) + \sum_{i \in I} \delta^*(b_i, x_i) \qquad (7.3)$$

Now, let us define the local quantities:

$$H_i(x) = -\sum_{j \in V_i} \delta(x_i, x_j) + \delta^*(b_i, x_i). \qquad (7.4)$$

Expression (7.4) may be seen as a local potential for each site on the graph G.

Let $\{f_i\}$ be a set of local rules on the sites of G,

$$f_i : Q^{|V_i|} \to Q \quad \text{for } i \in I$$

For symplicity of notation we write $f_i(x_j; j \in V_i)$ as $f_i(x)$.

We say that $\{f_i\}$ are *compatible* local functions iff:

$$\forall i \in I : \quad H_i(\tilde{x}) \leq H_i(x), \quad \forall x \in Q^n \qquad (7.5)$$

where $x = (x_1, ..., x_n)$, $\tilde{x} = (x_1, ..., x_{i-1}, f_i(x), x_{i+1}, ..., x_n)$.

That is to say, the application of each local rule decreases Hamiltonian (7.3).

In order to prove this, let us introduce:

$$a_{ij} = \begin{cases} 1 \text{ if } (i,j) \in V \\ 0 \text{ otherwise} \end{cases}$$

Clearly $A = (a_{ij})$ is an $n \times n$ symmetric matrix and, since $(i, i) \notin V$, $\mathrm{diag} A = 0$. Now, we may write H as follows:

$$H(x) = -\frac{1}{2} \sum_{i \in I} \sum_{j \in I} a_{ij} \delta(x_i, x_j) + \sum_{i \in I} \delta^*(b_i, x_i)$$

Let us assume that $\tilde{x} = (x_1, ..., f_k(x), ..., x_n)$. From the symmetry of A and δ, δ^* we may write:

$$\Delta H = H(\tilde{x}) - H(x) = -\frac{1}{2} \sum_{i \neq k} \sum_{j \neq k} a_{ij} \delta(x_i, x_j) - \sum_{j \in I} a_{kj} \delta(f_k(x), x_j)$$

$$+ \sum_{i \neq k} \delta^*(b_i, x_i) + \delta^*(b_k, f_k(x)) + \frac{1}{2} \sum_{i \neq k} \sum_{j \neq k} a_{ij} \delta(x_i, x_j)$$

$$+ \sum_{j \in I} a_{kj} \delta(x_k, x_j) - \sum_{i \neq k} \delta^*(b_i, x_i) - \delta^*(b_k, x_k)$$

then

$$\Delta H = - \sum_{j \in V_k} \delta(f_k(x), x_j) + \delta^*(b_k, f_k(x)) + \sum_{j \in V_k} \delta(x_k, x_j) - \delta^*(b_k, x_k)$$

$$= H_k(\tilde{x}) - H_k(x) \leq 0$$

which in non-positive, as seen directly from definition (7.5).

The above Compatible Networks may be seen as local strategies to minimize functional (7.3). Obviously these strategies, because of their local nature, lead only to local minima of (7.3). In this context (7.3) is a Lyapunov functional for the sequential update:

    0.   Take $x \in Q^n$

    1.   Repeat until reaching a steady state   (7.6)

        For $i \leftarrow 1, ..., n : x_i \leftarrow f_i(x)$.

Obviously scheme (7.6) is equivalent to the sequential one presented in previous chapters. We have the following result:

Proposition 7.1. Given a Compatible Network, Hamiltonian (7.3) is a Lyapunov functional for any sequential trajectory $\{x(t)\}_{t \geq 0}$.

Proof. Direct from the above remarks. ∎

It is important to point out that (7.3) is not necessarily strictly decreasing i.e., we may have $f_i(x) \neq x_i$ but $H_i(\tilde{x}) = H_i(x)$. We may avoid this problem by taking *strictly compatible* functions i.e.: $H_i(\tilde{x}) = H_i(x)$ iff $f_i(x) = x_i$. Clearly in this case we insure convergence to fixed points i.e., local minima of H. In several applications we do not need the strict hypothesis to prove convergence. In such cases we add an external magnetic field ad-hoc to get a strictly decreasing operator in the transient phase.

As examples of Compatible Networks we have:

7.2.1. Majority Networks. Let $\delta = \delta_K$ the Kroeneker function, $\delta^* = 0$ and $Q = \{0, ..., q-1\}$ be the state set. For $i \in I$ define:

$$f_i(x) = s \in Q \iff |\{j \in V_i/x_j = s\}| \geq |\{j \in V_i/x_j = r\}| \text{ for any } r \in Q$$

and in case of a tie the maximum value is chosen.

This function may be written, in a more compact form, as follows:

$$f_i(x) = \max\{s: \quad |\{j \in V_i/x_j = s\}| \geq \max_{0 \leq l \leq q-1} |\{j \in V_i/x_j = l\}| \} \tag{7.7}$$

The foregoing local rules $\{f_i\}$ are compatible with the Potts Hamiltonian (7.1), $H(x) = -\frac{1}{2} \sum_{(i,j) \in V} \delta_K(x_i, x_j)$. In fact by taking $\tilde{x} = (...x_{i-1}, f_i(x), x_{i+1}...)$ we get:

$$(\Delta H)_i = H_i(\tilde{x}) - H_i(x)$$

$$= -\sum_{j \in V_i} \delta_K(f_i(x), x_j) + \sum_{j \in V_i} \delta_K(x_i, x_j)$$

$$= -|\{j \in V_i/x_j = f_i(x)\}| + |\{j \in V_i/x_j = x_i\}| \leq 0$$

which is non-increasing, as found from the definition of f_i.

7.2.2. Next Majority Rule. By taking always the Kroeneker function δ_K, $\delta^* = 0$, and $Q = \{0, ..., q-1\}$ one defines:

$$f_i(x) = \begin{cases} x_i + 1(\text{mod}q) & \text{iff } |\{j \in V_i/x_j = x_i + 1(\text{mod}q)\}| \geq |\{j \in V_i/x_j = x_i\}| \\ x_i & \text{otherwise} \end{cases}$$

$$\tag{7.8}$$

we may see this rule with the scheme

$$0 \rightarrow 1 \rightarrow 2 \rightarrow ... \rightarrow q-1 \rightarrow 0$$

where the change from s to $s + 1(\mathrm{mod}q)$ occurs if and only if the next possible state is represented in the neighborhood at least as the current state s.

As in example 7.2.1 this rule is compatible i.e.:

$$(\Delta H)_i = H_i(\tilde{x}) - H_i(x) = -|\{j \in V_i / x_j = f_i(x)\}| + |\{j \in V_i / x_j = x_i\}|$$
$$\text{since } f_i(x) \in \{x_i, x_i + 1(\mathrm{mod}q)\} \text{ we have } (\Delta H)_i \leq 0.$$

If we replace in (7.8) the $\leq$ by $<$ we get $x_i \neq f_i(x) \Rightarrow H_i(\tilde{x}) < H_i(x)$, hence this new rule is strictly compatible, and therefore the sequential iteration converges to fixed points.

7.2.3. Median Rule. Let us take $Q = \{-M, ..., 0, ..., +M\}$, $\delta(a,b) = \min(a,b)$, $\delta^*(a,b) = ab$, $d_i = |V_i|$, and the local rules:

$$f_i(x) = \begin{cases} x_i - 1 & \text{if} \quad |\{j \in V_i / x_j < x_k\}| > \frac{d_i}{2} \\ x_i + 1 & \text{if} \quad |\{j \in V_i / x_j > x_k\}| > \frac{d_i}{2} \\ x_i & \text{otherwise} \end{cases} \tag{7.9}$$

This rule is well defined since if one of the cardinal is greater than $\frac{d_i}{2}$ necessarily the other is smaller. It is a kind of median local rule because the next state is the median local value of its neighbors. For instance:

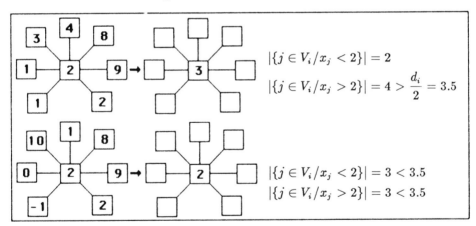

$$|\{j \in V_i / x_j < 2\}| = 2$$
$$|\{j \in V_i / x_j > 2\}| = 4 > \frac{d_i}{2} = 3.5$$

$$|\{j \in V_i / x_j < 2\}| = 3 < 3.5$$
$$|\{j \in V_i / x_j > 2\}| = 3 < 3.5$$

Figure 7.1. Local evolution of Next Majority rule.

Here the generalized Potts Hamiltonian is:

$$H(x) = -\frac{1}{2} \sum_{(i,j) \in V} \min(x_i, x_j) + \sum_{i \in I} \frac{d_i}{2} x_i \tag{7.10}$$

Lemma 7.1. Rule (7.9) is strictly compatible with Hamiltonian (7.10).

Proof. Let us update the i-th site, we have from the local potential (7.4):

$$(\Delta H)_i = H_i(\tilde{x}) - H_i(x) = \sum_{j \in V_i} (-\min(f_i(x), x_j) + \min(x_i, x_j)) + \frac{d_i}{2}(f_i(x) - x_i)$$

Clearly $f_i(x) = x_i \Rightarrow (\Delta H)_i = 0$. Let us suppose then $f_i(x) \neq x_i$. First we take $f_i(x) = x_i + 1$, hence:

$$(\Delta H)_i = \sum_{j \in V_i} (-x_i - 1 + x_i) + \frac{d_i}{2}$$

$$= -|\{j \in V_i / x_j > x_i\}| + \frac{d_i}{2}$$

Since $f_i(x) = x_i + 1$ iff $|\{j \in V_i / x_j > x_i\}| > \frac{d_i}{2}$ we get $(\Delta H)_i < 0$. Similarly, if $f_i(x) = x_i - 1$ we obtain

$$(\Delta H)_i = |\{j \in V_i / x_j \geq x_i\}| - \frac{d_i}{2}$$

$$\text{since } |\{j \in V_i / x_j < x_i\}| > \frac{d_i}{2} \quad \text{we get } (\Delta H)_i < 0$$

Hence the $\{f_i\}$ are strictly compatible local rules. ∎

Clearly, since the above rules are strict, the sequential update converges to fixed points that are local minima of (7.10).

7.2.4. Threshold Functions. Let $\delta(a, b) = \delta^*(a, b) = ab$ and the Hamiltonian:

$$H(x) = -\frac{1}{2} \sum_{(i,j) \in V} x_i x_j + \sum_{i \in I} b_i x_i \tag{7.11}$$

Recall that (7.11) is the classical Ising Hamiltonian [GV,W]. We define, for $i \in I$:

$$f_i(x) = \begin{cases} 0 & \text{if } \sum_{j \in V_i} x_j - b_i < 0 \\ 1 & \text{otherwise} \end{cases}$$

The above rules are compatible. In fact, by updating the i-th site:

$$(\Delta H)_i = -\sum_{j \in V_i} f_i(x)x_j + b_i f_i(x) + \sum_{j \in V_i} x_i x_j - b_i x_i$$

$$= -(f_i(x) - x_i)(\sum_{j \in V_i} x_j - b_i) \leq 0$$

Threshold functions define the Neural Network model studied in previous chapters. Here we have the particular case with weights $a_{ij} \in \{0,1\}$. Furthermore, functional H is in this case the Ising Hamiltonian.

As we saw in previous chapters, if for any $i \in I$, and $x \in \{0,1\}^n$
$\sum_{j \in V_i} x_j - b_i \neq 0$ (for instance $b_i \in Q \setminus \mathbb{Z}$), then the threshold functions are strictly compatible. Hence the sequential iteration admits only fixed points which are local minima of the Ising Hamiltonian (7.11).

7.3. The Complexity of Synchronous Iteration on Compatible Rules

As we have seen in previous sections, the sequential update on Compatible Networks admits H as a Lyapunov functional which usually drives the network dynamics to fixed points, i.e. we obtain a simple dynamic behaviour. The question now is what happens when, for Compatible Networks, we change the iteration mode, for instance by using a synchronous update. The answer is not simple, but for some classes it is possible to establish, from the Hamiltonian H, a Lyapunov operator driving the synchronous dynamics which may be characterized in the steady state by a two-periodic behaviour (fixed points and/or two-cycles). On the other hand, other Compatible Networks have a complex synchronous dynamics which is measured as the computing capabilities of the network. More precisely, we prove that there exist Compatible Networks capable to calculate, by coding bits as configurations in the network, any logic function and one-dimensional compatible Cellular Automata which simulate a Universal Turing Machine.

7.3.1. Logic Calculator. Given a set $\{f_i\}$ of compatible functions with a Hamiltonian H, the synchronous iteration is given by the scheme:

$$x_i(t+1) = f_i(x(t)) \quad i \in I, \ x(0) \in Q^n.$$

Let $G = (\mathbb{Z}, V_0 = \{-1,1\})$, so the neighborhood of each $i \in \mathbb{Z}$ is $V_i = \{i-1, i+1\}$. Also take $Q = \{0,1,2,3,\}$ and the Next Majority rule introduced in example 2 of section 7.2:

$$f_i(x_{i-1}, x_{i+1}) = \begin{cases} x_i + 1 (\mathrm{mod}4) \text{ if } |\{j \in V_i / x_j = x_i + 1 (\mathrm{mod}4)\}| \\ \qquad\qquad \geq |\{j \in V_i / x_j = x_i\}| \\ x_i \text{ otherwise} \end{cases} \qquad (7.12)$$

Since $\mathbb{Z}$ is an infinite set, $H(x)$ diverges for non finite $x \in Q^{\mathbb{Z}}$. But, as we are only interested in finite evolutions of the automaton, we will always take a finite set $\{0, ..., n-1\} \subset \mathbb{Z}$ large enough to support the computations described below.

The above function updated synchronously admits complex dynamical configurations, some of which are the gliders exhibited in Figure 7.2.

$$
\begin{array}{ccccccc}
0 & 3 & 2 & 1 & 0 & 0 & 0 \\
0 & 0 & 3 & 2 & 1 & 0 & 0 \\
0 & 0 & 0 & 3 & 2 & 1 & 0
\end{array}
\qquad
\begin{array}{l}
t \\
t+1 \quad \text{right vehicle} \\
t+2
\end{array}
$$

$$
\begin{array}{ccccccc}
0 & 0 & 0 & 1 & 2 & 3 & 0 \\
0 & 0 & 1 & 2 & 3 & 0 & 0 \\
0 & 1 & 2 & 3 & 0 & 0 & 0
\end{array}
\qquad
\begin{array}{l}
t \\
t+1 \quad \text{left vehicle} \\
t+2
\end{array}
$$

Figure 7.2. Gliders: Configurations move in the lattice breaking the symmetry of the cellular space.

The interaction between two gliders (crash of gliders) leads to the quiescent configuration, as shown in Figure 7.3.

$$
\begin{array}{ccccccccccc}
0 & 3 & 2 & 1 & 0 & 0 & 0 & 1 & 2 & 3 & 0 \\
0 & 0 & 3 & 2 & 1 & 0 & 1 & 2 & 3 & 0 & 0 \\
0 & 0 & 0 & 3 & 2 & 1 & 2 & 3 & 0 & 0 & 0 \\
0 & 0 & 0 & 0 & 3 & 2 & 3 & 0 & 0 & 0 & 0 \\
0 & 0 & 0 & 0 & 0 & 3 & 0 & 0 & 0 & 0 & 0 \\
0 & 0 & 0 & 0 & 0 & 0 & 0 & 0 & 0 & 0 & 0
\end{array}
\qquad
\begin{array}{l}
t = 0 \\
1 \\
2 \quad \text{odd meeting} \\
3 \\
4 \\
5
\end{array}
$$

$$
\begin{array}{cccccccc}
3 & 2 & 1 & 0 & 0 & 1 & 2 & 3 \\
0 & 3 & 2 & 1 & 1 & 2 & 3 & 0 \\
0 & 0 & 3 & 2 & 2 & 3 & 0 & 0 \\
0 & 0 & 0 & 3 & 3 & 0 & 0 & 0 \\
0 & 0 & 0 & 0 & 0 & 0 & 0 & 0
\end{array}
\qquad
\begin{array}{l}
t = 0 \\
1 \\
2 \quad \text{even meeting} \\
3 \\
4
\end{array}
$$

Figure 7.3. Interaction between gliders.

It is easy to see that finite vectors $(0^* 2^s 0^*)$ where $s \geq 2$ are stable configurations, i.e. the application of synchronous iteration on such a configuration is invariant, i.e. $(0^* 2^s 0^*) \rightarrow (0^* 2^s 0^*)$.

The interaction between gliders and stable configurations (by crashing) erases the stable configuration and leads to the quiescent state 0, as in the evolution in Figure 7.4.

```
3  2  1  0  0  2  2  2  0        3  2  1  0  2  2  0  1  2  3
0  3  2  1  0  2  2  2  0        0  3  2  1  2  2  1  2  3  0
0  0  3  2  1  2  2  2  0        0  0  3  2  2  2  2  3  0  0
0  0  0  3  2  2  2  2  0        0  0  0  3  2  2  3  0  0  0
0  0  0  0  3  2  2  2  0        0  0  0  0  3  3  0  0  0  0
0  0  0  0  0  3  2  2  0        0  0  0  0  0  0  0  0  0  0
0  0  0  0  0  0  3  2  0
0  0  0  0  0  0  0  3  0
0  0  0  0  0  0  0  0  0
```

Figure 7.4. Interaction between gliders and stable configurations.

By using previous configurations we may simulate logic gates by coding them as gliders. In fact we may simulate the AND, NOR, and OR gates in the following way:

NOR-gate. $u \to \bar{u}$; $u \in \{0,1\}$; $\bar{u} = 1 - u$

$$\cdots 0321\text{XXX}\underline{000}\cdots$$
$$\text{output cells}$$

where $\text{XXX} \in \{123, 000\}$ is the code of the input variable u.

In six steps the output, consisting of the three labeled sites, gives the negation of the input variable.

OR-gate. $(u,v) \to u \vee v$; $u, v \in \{0,1\}$

$$\cdots 032100\text{XXX}0220\text{YYY}\underline{000}\cdots$$
$$\text{output}$$

$\text{XXX} \in \{000, 321\}$, $\text{YYY} \in \{000, 123\}$, where XXX codes u and YYY codes v.

AND-gate: $(u,v) \to u \wedge v$; $u, v \in \{0,1\}$

$$\cdots 032100\text{XXX}00\text{YYY}0022000000000000123\underline{000}\cdots$$
$$\text{output}$$

$\text{XXX}, \text{YYY} \in \{000, 321\}$.

```
(i)3210032100321002200000000000000123
   0321003210032102200000000000001230
   0032100321003212200000000000012300
   0003210032100322200000000000123000
   0000321003210032200000000001230000
   0000032100321003200000000012300000
   0000003210032100300000000123000000
   0000000321003210000000012300000000
   0000000032100321000000123000000000
   0000000003210032100012300000000000
   0000000000321003210123000000000000
   0000000000032100321230000000000000
   0000000000003210032300000000000000
   0000000000000321003000000000000000
   0000000000000032100000000000000000
   0000000000000003210000000000000000
   0000000000000000321000000000000000
   0000000000000000032100000000000000
   0000000000000000003210000000000000
   0000000000000000000321000000000000
   0000000000000000000032100000000000
   0000000000000000000003210000000000
   0000000000000000000000321000000000
   0000000000000000000000032100000000
   0000000000000000000000003210000000
   0000000000000000000000000321000000
   0000000000000000000000000032100000
   0000000000000000000000000003210000
   0000000000000000000000000000321000
   0000000000000000000000000000032100
   0000000000000000000000000000003210
   0000000000000000000000000000000321
(ii)3210032100000002200000000000000123
   0321003210000002200000000000001230
   0032100321000002200000000000012300
   0003210032100002200000000000123000
   0000321003210002200000000001230000
   0000032100321002200000000012300000
   0000003210032102200000001230000000
   0000000321003212200000123000000000
   0000000032100322200012300000000000
   0000000003210032200123000000000000
   0000000000321003200123000000000000
   0000000000032100301230000000000000
   0000000000003210012300000000000000
   0000000000000321012300000000000000
   0000000000000032123000000000000000
   0000000000000003230000000000000000
   0000000000000003000000000000000000
```

```
(iii)000321123000
     000032230000
     000003300000
     000000000000

(iv)000321000000
    000032100000
    000003210000
    000000321000
    000000032100
    000000003210
    000000000321

(iv)3210032100220000000
    0321003212200000000
    0032100322200000000
    0003210032200000000
    0000321003200000000
    0000032100300000000
    0000003210000000000
    0000000321000000000
    0000000032100000000
    0000000003210000000
    0000000000321000000
    0000000000032100000
    0000000000003210000
    0000000000000321000
    0000000000000032100
    0000000000000003210
    0000000000000000321

(v)321000000220123000
   032100000221230000
   003210000222300000
   000321000223000000
   000032100230000000
   000003210300000000
   000000321000000000
   000000032100000000
   000000003210000000
   000000000321000000
   000000000032100000
   000000000003210000
   000000000000321000
   000000000000032100
   000000000000003210
```

Figure 7.5. Dynamics of logical gates.

Figures 7.5.i and 7.5.ii show the dynamics of the AND-gate for $(u,v) = (1,1)$ and $(u,v) = (1,0)$ respectively. Figures 7.5.iii and 7.5.iv exhibit the NOR-gate for $u = 1$ and $u = 0$ respectively. Finally, Figures 7.5.iv and 7.5.v show the dynamic for the OR-gate in the cases $(u,v) = (1,0)$ and $(u,v) = (0,1)$ respectively.

Since the set {NOR, OR, AND} is Universal (i.e. any logical function may be written with these operators), with these gates the network may simulate any logical function. Hence, in terms of computing capabilities, the automaton is complex, in spite of the fact that it is compatible.

As a last example, we give the code for the 3-variable logic function $\varphi = (a \vee b) \wedge \bar{c}$:

$$\cdots 03210^2 3210AAA022BBB0^2 321CCC00220^{26} 123\underline{000} \cdots$$

output

where $v^s = v...v$, s times; and $AAA \in \{000, 321\}$, $BBB, CCC \in \{000, 123\}$ code the values a, b, c respectively. The code is explained in Figure 7.6.

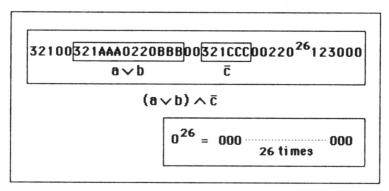

Figure 7.6. Configuration which simulates $(a \vee b) \wedge \bar{c}$.

Recall that the last one-dimensional automaton accepts non-bounded cycles. It suffices to take a one-dimensional n-torus and a glider which realizes a shift of length n.

In the two-dimensional Next Majority Automaton $\mathcal{A} = (\{0,1,2,3\}, V_0^N, f)$, typical patterns are as in Figure 7.7 and 7.8. Some of the patterns observed are very complex; for instance, in Figure 7.7 periodic non-bounded patterns appear which clearly increase with the torus size.

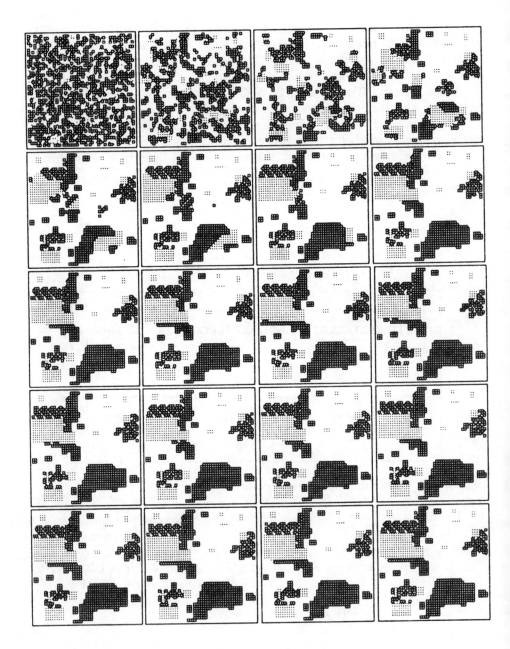

Figure 7.7. Synchronous update of the Next Majority Rule in a 40×40 torus. Steps are given from left to right and from top to bottom.

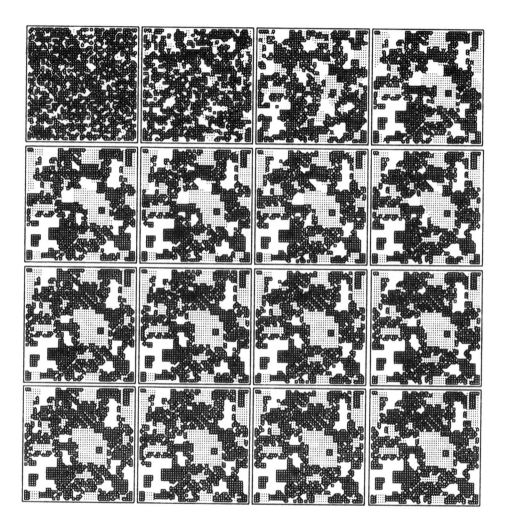

Figure 7.8. Synchronous dynamics of the Next Majority Rule in a 40×40 torus. Steps are given from left to right and from top to bottom.

7.3.2. Potts Universal Automaton. Now we shall prove that a Universal Turing Machine may be simulated by a one-dimensional Potts Automata. We shall follow the notations and results introduced in Chapter 1.

Let $T = (Q, \Sigma, g)$ be a Turing Machine. Q is the finite set of internal states, Σ is the alphabet or input output set with a particular symbol, the "blank" is denoted "0" and $g : Q \times \Sigma \to Q \times \Sigma \times \{-1, 0, 1\}$ is the transition function which associates to any pair (q, s) of an internal state and a scanned symbol a triple (q', s', d') which contains the new internal state, the printed symbol, and the moving direction respectively.

Proposition 7.2. Given $T = (Q, \Sigma, g)$ there exists a Potts Automaton whose synchronous dynamics simulates T.

Proof. We take the one-dimensional automaton $A = (\mathbb{Z}, V_0, Q^*, f)$ where $V_0 = \{-1, 0, 1\}$ is the neighbourhood, $Q^* = (Q \cup \{*\}) \times \Sigma$ is the set of states, with $(*, 0)$ being its quiescent state.

The local function is defined as in Chapter 1,

$$f((q, s), (*, u), (*, v)) = \begin{cases} (q', u) \text{ if } g(q, s) = (q', s', +1) \\ (*, u) \text{ otherwise} \end{cases}$$

$$f((*, u), (q, s), (*, v)) = \begin{cases} (*, s') \text{ if } g(q, s) = (q', s' \pm 1) \\ (q', s') \text{ if } g(q, s) = (q', s, ,0) \end{cases}$$

$$f((*, u), (*, v), (q, s)) = \begin{cases} (q', v) \text{ if } g(q, s) = (q', s', -1) \\ (*, u) \text{ otherwise} \end{cases}$$

For any other local configuration $f(x, y, z) = y \in Q^*$.

Clearly, as we saw in Chapter 1, A simulates T. To determine a Potts Hamiltonian associated to the network define the following symmetric function δ:
For $s, s' \in \Sigma$, $q, q' \in Q$:

$$\delta((q, s), (q', s')) = b \in \mathbb{N}$$

For $s, s' \in \Sigma \setminus \{0\}$:

$$\delta((*, s), (*, s')) = a \in \mathbb{N}$$

For $s, s' \in \Sigma$, $q \in Q$:

$$\delta((*, s), (q, s')) = a \in \mathbb{N}$$

where $0 < a < b$ and $\delta((*, 0), (*, 0)) = 0$.

Now we shall define the functional H of the Potts automaton. For any configuration x of finite support, let

$$H(x) = -\frac{1}{2} \sum_{(i,j) \in V} \delta(x_i, x_j)$$

hence the local potentials are:

$$H_i(x) = -\delta(x_i, x_{i-1}) - \delta(x_i, x_{i+1}) \quad \forall i \in \mathbb{Z}$$

Clearly f is compatible. In fact, let us update the site $i \in \mathbb{Z}$, the only cases where $\tilde{x} \neq x$ are the following:

$$\downarrow^i$$

1. $$x = ...(q,s)(*,u)(*,v)...$$
$$\text{and } \tilde{x} = ...(q,s)(q',u)(*,v)...$$

$$\downarrow^i$$

2. $$x = ...(*,u)(*,v)(q,s)...$$
$$\text{and } \tilde{x} = ...(*,u)(q',v)(q,s)...$$

$$\downarrow^i$$

3. $$x = ...(*,u)(q,s)(*,v)...$$
$$\text{and } \tilde{x} = ...(*,u)(*,s')(*,v)$$
$$\text{or } \tilde{y} = ...(*,u)(q',s')(*,v)$$

The cases 1 and 2 are similar, it suffices to analyze case 1. We have:

$$H_i(x) = -2a \text{ and } H_i(\tilde{x}) = -b - a \quad \text{hence } H_i(\tilde{x}) < H_i(x).$$

In both situations of case 3 we get:

$$H_i(x) = -2a, \ H_i(\tilde{x}) = -2a, \ H_i(\tilde{y}) = -2a$$

So $H_i(x) = H_i(\tilde{x}) = H_i(\tilde{y})$. Then f is a compatible function. ∎

Corollary 7.1. There exists a Universal one-dimensional Potts Automaton.

Proof. Direct from Proposition 7.2 and the results of Chapter 1. ∎

Remarks. The function δ defined in the theorem gives a non-trivial Potts Automaton, because in some cases the local application of f diminishes strictly the energy. Also, the sequential application of f usually leads to fixed points, hence we have a qualitative change in the dynamical complexity: the simpler dynamics occurs in sequential iterations (fixed points) and the higher complexity in synchronous updating (simulation of a Universal Turing Machine).

7.4. Solvable Classes for the Synchronous Update

Here we present two classes of Compatible Networks which accept Lyapunov operators driving their synchronous dynamics. In spite of their simple behaviour (fixed points or two-cycles) these classes are important in image-restoration, majority iterations in population dynamics, etc. [GMR,PoV]. Some of these applications will be presented below.

7.4.1. Maximal Rules. Given the Hamiltonian (7.3) we define the *Maximal Compatible* local rules as follows:

$$\forall\, i = 1,...,n: \quad f_i(x) = s \iff \sum_{j \in V_i} \delta(s, x_j) - \delta^*(b_i, s) \geq$$

$$\geq \sum_{j \in V_i} \delta(r, x_j) - \delta^*(b_i, r) \quad \forall r \in Q \tag{7.13}$$

in case of a tie, the maximum state is taken.

Clearly, rule (7.13) is compatible with local potential (7.4). In fact, from definition (7.13) we get:

$$(\Delta H)_i = H_i(\tilde{x}) - H_i(x) = -\sum_{j \in V_i} \delta(f_i(x), x_j) - \delta^*(b_i, f_i(x))$$

$$+ \sum_{j \in V_i} \delta(x_i, x_j) + \delta^*(b_i, x_i) \leq 0$$

Rule (7.13) is a local steepest descent; i.e. it is a rule that causes the Hamiltonian to decrease most among the local compatible configurations.

Theorem 7.1. For a non-oriented graph G and a set $\{f_i\}$ of maximal local rules, any synchronous trajectory $\{x(t)\}_{t \geq 0}$ accepts the following quantity as a Lyapunov functional:

$$E_{sy}(x(t)) = -\sum_{(i,j) \in V} \delta(x_i(t), x_j(t-1)) + \sum_{i \in I}(\delta^*(b_i, x_i(t)) + \delta^*(b_i, x_i(t-1))) \tag{7.14}$$

Proof. Let us take $\Delta_t E_{sy} = E_{sy}(x(t)) - E_{sy}(x(t-1))$. Since G is non oriented and δ is symmetric:

$$\Delta_t E_{sy} = \sum_{i \in I} \{-\sum_{j \in V_i} \delta(x_i(t), x_j(t-1)) + \delta^*(b_i, x_i(t))$$
$$+ \sum_{j \in V_i} \delta(x_i(t-2), x_j(t-1)) - \delta^*(b_i, x_i(t-2))\}$$

From the definition of f_i, each term in the above sum satisfies $(\Delta_t E_{sy})_i \leq 0$, hence $E_{sy}(x(t)) \leq E_{sy}(x(t-1)) \ \forall t \geq 1.$ ∎

It is direct to note that, in case of a tie we may have $x_i(t) \neq x_i(t-2)$ and $(\Delta_t E_{sy})_i = 0$. That is to say, the network evolves from $x(t-2)$ to $x(t)$, which are different states, but the functional E_{sy} does not account for it. In order to avoid this, one takes a perturbation of E_{sy} to obtain a strictly decreasing quantity in the transient phase. Then define:

$$E_{sy}^*(x(t)) = E_{sy}(x(t)) - \frac{\epsilon}{q} \sum_{i \in I} (x_i(t) + x_i(t-1)) \tag{7.15}$$

where $q = |Q|$ and ϵ is the smallest positive value taken by the quantity $\min_i |(\Delta E_{sy})_i|$.

Theorem 7.2. Under the preceding hypothesis $\{E_{sy}^*(x(t))\}_{t \geq 1}$ is a strictly decreasing Lyapunov functional.

Proof. We have

$$\Delta_t E_{sy}^* = \Delta_t E_{sy} - \frac{\epsilon}{q} \sum_{i \in I} (x_i(t) - x_i(t-2)).$$

Consider the tie case: $x_i(t) \neq x_i(t-2)$ and $(\Delta_t E_{sy})_i = 0$. From the definition of f_i we get $x_i(t) > x_i(t-2)$, hence:

$$(\Delta_t E_{sy}^*)_i = -\frac{\epsilon}{q}(x_i(t) - x_i(t-2) \leq -\frac{\epsilon}{q} < 0.$$

If $x_i(t) \neq x_i(t-2)$ and $(\Delta_t E_{sy})_i < 0$ then $(\Delta_t E_{sy})_i \leq -\epsilon$ so:

$$(\Delta_t E_{sy}^*)_i \leq -\epsilon - \frac{\epsilon}{q}(1-q) = -\frac{\epsilon}{q} < 0$$ ∎

For the sequential iteration case it is easy to see that for any sequential trajectory $\{x(t)\}_{t \geq 0}$ the quantity:

$$E^*_{seq}(x(t)) = H(x(t)) - \frac{\epsilon}{q} \sum_{i \in I} x_i(t) \qquad (7.16)$$

is a strictly decreasing Lyapunov functional. In fact, if we update the i-th cell we get:

$$\Delta E^*_{seq} = - \sum_{j \in V_i} \delta(f_i(x), x_j) + \delta^*(f_i(x), x_j)$$

$$+ \sum_{j \in V_i} \delta(x_i, x_j) - \delta^*(x_i, x_j) - \frac{\epsilon}{q}(f_i(x) - x_i)$$

Since f_i is maximal if $f_i(x) \neq x_i$ and in a non-tie case:

$$\Delta E^*_{seq} = -\epsilon - \frac{\epsilon}{q}(f_i(x) - x_i) \leq -\frac{\epsilon}{q} < 0$$

in case of a tie:

$$\Delta E^*_{seq} = -\frac{\epsilon}{q}(f_i(x) - x_i) \leq -\frac{\epsilon}{q} < 0.$$

From the previous theorem and the remark above we may state:

Theorem 7.3. Given a non-oriented graph G and a set $\{f_i\}$ of maximal compatible functions, the sequential iteration converges to fixed points and the synchronous iteration to fixed points and/or two cycles.

Proof. Direct from Theorem 7.2 and the above remark by using E^*_{seq} and E^*_{sy}.
∎

As examples of Maximal Local Rules we have the following ones:

7.4.1.1. Majority Networks. Here we take the local rule of example 7.2.1, i.e. $\delta = \delta_K$ (the Kroeneker function), $\delta^* = 0$ and f_i is the rule defined in (7.7). It is not difficult to prove that the functions $\{f_i\}$ are maximal. In fact, it suffices to prove that the local rules may be written as

$$f_i(x) = s \iff \sum_{j \in V_i} \delta_K(s, x_j) \geq \sum_{j \in V_i} \delta_K(r, x_j), \quad \forall r \in Q \qquad (7.17)$$

and s is the maximum value in case of a tie.

The equivalence between (7.7) and (7.17) is obtained directly from the equality:

$$\sum_{j \in V_i} \delta_K\left(r, x_j\right) = \left|\{j \in V_i \,|\, x_j = r\}\right|$$

Proposition 7.3. The Majority Network admits E^*_{seq} and E^*_{sy} as strict Lyapunov functionals for the sequential and synchronous iteration respectively:

$$E^*_{seq}(x(t)) = -\frac{1}{2} \sum_{(i,j) \in V} \delta_K\left(x_i(t), x_j(t)\right) - \frac{1}{q} \sum_{i \in I} x_i(t)$$

$$E^*_{sy}(x(t)) = - \sum_{(i,j) \in V} \delta_k\left(x_i(t), x_j(t-1)\right) - \frac{1}{q} \sum_{i \in I}\left(x_i(t) + x_i(t+1)\right)$$

Then, in steady state, they have respectively only fixed points or the two cycle behaviour.

Proof. Direct from Theorem 7.3. It suffices to observe that the minimum difference in a non-tie case is $\epsilon = 1$. ∎

The majority rule is clearly non-strictly compatible which is the reason to add an internal magnetic field $-\frac{1}{q} \sum x_i$. For instance we may have the situation of Figure 7.9.

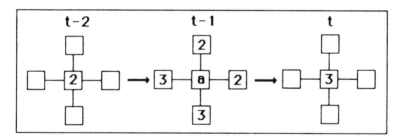

Figure 7.9. Local evolution of majority for $a \in Q$.

Hence locally:

$$\sum_{j \in V_i} \delta_K\left(x_i(t), x_j(t-1)\right) - \sum_{j \in V_i} \delta_K\left(x_i(t-2), x_j(t-1)\right) = 2 - 2 = 0 \text{ (a tie case)}$$

then $(Ex(t))_i - (Ex(t-1))_i = 0$

7.4.1.2. Local Coloring Rules. Let $\delta(u,v) = 0$ if $u = v$ and 1 if $u \neq v$. Take $Q = \{0, ..., q-1\}$ as the set of colors and consider the local rule:

$$f_i(x) = s \iff |\{j \in V_i | x_j \neq s\}| \geq |\{j \in V_i | x_j \neq r\}| \quad \forall r \in Q \qquad (7.18)$$

in case of a tie, the maximum is taken.

Clearly, (7.18) is equivalent to formulation (7.13) because:

$$\sum_{j \in V_i} \delta(s, x_j) = |\{j \in V_i | x_j \neq s\}|$$

As in the previous proposition we obtain:

Proposition 7.4. The local coloring rule (7.18) admits E^*_{seq}, E^*_{sy} as strict Lyapunov functionals where:

$$E^*_{seq}(x(t)) = -\frac{1}{2} \sum_{(i,j) \in V} \delta(x_i(t), x_j(t)) - \frac{1}{q} \sum_{i \in I} x_i(t)$$

$$E^*_{sy}(x(t)) = - \sum_{(i,j) \in V} \delta(x_i(t), x_j(t-1)) - \frac{1}{q} \sum_{i \in I} (x_i(t) + x_i(t-1))$$

Hence the sequential iteration admits only fixed points, while the synchronous update admits fixed points and/or two-cycles.

Proof. Direct from Theorem 7.2 and 7.3, with $\epsilon = 1$. ∎

The preceding network may be seen as a strategy for coloring a graph with the usual restriction; i.e. two neighbouring nodes $((i,j) \in V)$ must have a different color. In this way, the sequential updating is a kind of greedy-algorithm. For instance, let us take the graph of Figure 7.10:

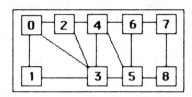

Figure 7.10. Non-oriented graph with labelled sites.

The respective evolutions of the sequential update in the order $0 < 1 < ... < 8$ and of the synchronous update for the same initial configuration are given in Figure 7.11.i and 7.11.ii.

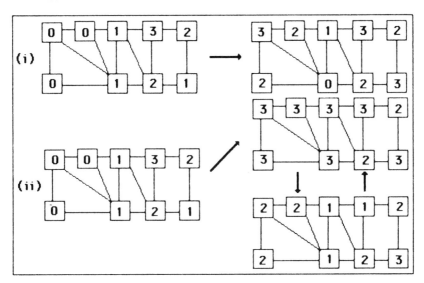

Figure 7.11. (i) Sequential evolution. (ii) Synchronous evolution.

In Figure 7.11.i the sequential iteration converges, from $x(0)$, to a fixed point which is a good coloration of G, but the synchronous update may be a bad strategy. In fact in Figure 7.11.ii it converges to a two-cycle which is not a solution of the color problem.

A variation of the above coloring rule is the following:

$$f_i(x) = x_i \qquad \text{iff } |\{j \in V_i | x_j \neq x_i\}| \geq |\{j \in V_i | x_j \neq r\}|, \ \forall r \in Q$$

$$\text{else:}$$

$$\begin{cases} s & \text{iff } |\{j \in V_i | x_j \neq s\}| \geq |\{j \in V_i | x_j \neq r\}|, \ \forall r \in Q \qquad (7.19) \\ \\ \text{and in case of a tie, the maximum } s \text{ is taken.} \end{cases}$$

Rule (7.19) may be interpreted as follows: the i-th site remains unchanged when any other color increases the local matching. Otherwise, it takes the maximum state among those which are at least as well represented in the neighbourhood. Clearly rule (7.19) is also maximal compatible for the same δ function:

If $f_i(x) = x_i$

$$\sum_{j \in V_i} \delta(f_i(x), x_j) - \sum_{j \in V_i} \delta(r, x_j) = |\{j \in V_i | x_j \neq x_i\}| - |\{j \in V_i | x_j \neq r\}| \geq 0$$

If $f_i(x) = s \neq x_i$ we get the same result.

Proposition 7.5. E_{sy} and E_{seq} are Lyapunov functionals for rule (7.19) where:

$$E_{seq}(x(t)) = -\frac{1}{2} \sum_{(i,j) \in V} \delta(x_i(t), x_j(t))$$

$$E_{sy}(x(t)) = - \sum_{(i,j) \in V} \delta(x_i(t), x_j(t-1))$$

Proof. In the sequential case the proof is direct from the fact that the rule is compatible, as is in the synchronous case because the rule is a maximal rule. ∎

More interesting is the following result:

Proposition 7.6. Any solution of the color problem is a fixed point of rule (7.19) for both iteration modes. Furthermore, the minimum value of the energy is attained for these configurations.

Proof. Since we know that the fixed points are the same for both iteration modes, it suffices to verify the second assertion for the synchronous one. Let $x \in Q^n$ be a solution of the color problem. For any $i \in I$, $j \in V_i$ we have $x_i \neq x_j$, so:

$$|\{j \in V_i / x_j \neq x_i\}| = |V_i| \geq |\{j \in V_i / x_j \neq r\}| \quad \forall r \in Q$$

Then $f_i(x) = x_i$ which is a fixed point.

On the other hand, the sequential energy is bounded by:

$$E_{seq}(u) = -\frac{1}{2} \sum_{(i,j) \in V} \delta(u_i, u_j) \geq -\frac{1}{2} \sum_{i \in I} |V_i|$$

hence $E_{seq}(u) \geq -\frac{1}{2} 2e = -e$, where e is the number of edges in the graph G.

Similarily, for any trajectory of a synchronous update

$$E_{sy}(x(t)) \geq -2e$$

Now if x is a solution of the color-problem we get:

$$E_{seq}(x) = -e \text{ and } E_{sy}(x) = -2e$$

the global minimum of both energy functionals. ∎

It is not difficult to see that function (7.18) does not fulfill the above properties; i.e. a solution of the color problem is not necessarily a fixed point. For instance consider the synchronous update on the graph $G = (I, V)$ where $I = \{1, 2, 3, 4, 5\}$, $V = \{(i,j) \in I \times I : |i - j| = 1\}$ and $Q = \{0, 1, 2, 3\}$ the set of colors. The configuration $x(0) = (1, 2, 1, 2, 1)$ is a solution of the color-problem but it is not a fixed point for rule (7.18):

$$
\begin{array}{ccccc}
1 & 2 & 1 & 2 & 1 \\
 & & \downarrow & & \\
3 & 3 & 3 & 3 & 3 \\
 & \downarrow & & \uparrow & \quad \text{two cycle} \\
2 & 2 & 2 & 2 & 2
\end{array}
$$

Figure 7.12. Synchronous iteration of rule (7.18).

Clearly, the two-cycle is not a solution of the color problem. Nevertheless, for the sequential update in the order $1 < 2 < 3 < 4 < 5$, $x(0)$ converges to a good solution (see Figure 7.13).

$$
\begin{array}{cccccl}
1^{\downarrow} & 2 & 1 & 2 & 1 & \\
3 & 2^{\downarrow} & 1 & 2 & 1 & \\
3 & 2 & 1^{\downarrow} & 2 & 1 & \\
3 & 2 & 3 & 2^{\downarrow} & 1 & \\
3 & 2 & 3 & 2 & 1^{\downarrow} & \\
3 & 2 & 3 & 2 & 3 & \text{fixed point}
\end{array}
$$

Figure 7.13. Sequential iteration of rule (7.18).

For the sequential update with rule (7.18) we have:

Lemma 7.2. If x is a solution of the color problem then any sequential trajectory of rule (7.18) with $x(0) = x$ belongs to the set of solutions of the color-problem.

Proof. Assume $f_i(x) = y \neq x_i$. Since x is a solution we get

$$|\{j \in V_i | x_j \neq y\}| = |\{j \in V_i / x_j \neq x_i\}| = |V_i|, \text{ hence the result} \quad ∎$$

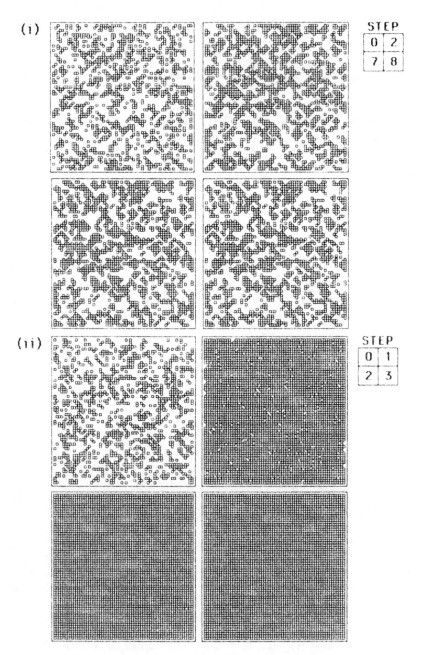

Figure 7.14. Dynamics of rule (7.18) for $|I| = 65$. Pixel $=$ "black" iff its color is different of all the neighbours colors. Synchronous (i) and sequential (ii) update.

In Figure 7.14 we compare the synchronous and the sequential update for the automaton $A = (I \times I, V_0^N, Q = \{0, 1, 2, 3\}, f)$, where V_0^N is the von-Neumann neighbourhood and f is rule (7.18). Any site belonging to $\mathbb{Z} \times \mathbb{Z} \setminus I \times I$ is assumed to be fixed in color 0. For random initial configurations the synchronous update evolves to a fixed point which is not a color solution (see the blank island in Figure 7.14.i). Nevertheless, the sequential update evolves to a color solution (see Figure 7.14.ii).

7.4.2. Smoothing Rules. For Hamiltonian (7.3) and the state space $Q = \{0, ..., q-1\}$ another class of compatible functions that accepts a Lyapunov functional in synchronous mode is the following one:

$$
f_i(x) = \begin{cases}
x_i - 1 \in Q & \text{if} \quad \sum_{j \in V_i} \delta(x_i - 1, x_j) - \delta^*(x_i - 1, b_i) > \\
& \qquad \sum_{j \in V_i} \delta(r, x_j) - \delta^*(r, b_i); \quad r = x_i, x_i + 1 \\
x_i + 1 \in Q & \text{if} \quad \sum_{j \in V_i} \delta(x_i + 1, x_j) - \delta^*(x_i + 1, b_i) > \\
& \qquad \sum_{j \in V_i} \delta(r, x_j) - \delta^*(r, b_i); \quad r = x_i, x_i - 1 \\
x_i & \text{if} \quad \text{either } x_i - 1, \ x_i + 1 \notin Q \\
& \qquad \text{or the above inequalities do not hold}
\end{cases}
\tag{7.20}
$$

Clearly rule (7.20) is compatible. Furthermore, for any synchronous trajectory $\{x(t)\}_{t \geq 0}$ we have $|x_i(t) - x_i(t-1)| \leq 1$ and $|x_i(t) - x_i(t-2)| \leq 2$. We obtain the following result:

Theorem 7.4. The quantities:

$$
E_{seq}(x(t)) = -\frac{1}{2} \sum_{(i,j) \in V} \delta(x_i(t), x_j(t)) + \sum_{i \in I} \delta^*(b_i, x_i(t))
$$

$$
E_{sy}(x(t)) = - \sum_{(i,j) \in V} \delta(x_i(t), x_j(t-1)) + \sum_{i \in I} (\delta^*(b_i, x_i(t)) + \delta^*(b_i, x_i(t-1)))
$$

are respectively strict Lyapunov functionals for the sequential and synchronous updating modes.

Proof. The sequential expression follows directly since the f_i are compatible. For a synchronous update:

$$
\Delta E_{sy} = - \sum_i \Big(\sum_{j \in V_i} \delta(x_i(t), x_j(t-1)) - \delta^*(b_i, x_i(t)) \\
- \sum_{j \in V_i} \delta(x_i(t-2), x_j(t-1)) + \delta^*(b_i, x_i(t-2)) \Big)
$$

Let us analyze the i-th term $(\Delta E_{sy})_i$ in the above sum:

$$\text{If} \quad x_i(t) = x_i(t-2) \quad \text{then} \ (\Delta E_{sy})_i = 0.$$

If $x_i(t) \neq x_i(t-2)$ and since $|x_i(t) - x_i(t-2)| \leq 2$, then two cases can occur:

(i) For $|x_i(t) - x_i(t-2)| = 1$ we may have:

$t-2$	$t-1$	t
a	a	$a+1$
a	$a+1$	$a+1$
a	a	$a-1$
a	$a-1$	$a-1$

Let us analyse the first case (the others are similar). From the definition of f_i:

$$\sum_{j \in V_i} \delta(a+1, x_j(t)) - \delta^*(a+1, x_i(t)) > \sum_{j \in V_i} \delta(a, x_j(t)) - \delta^*(a, x_i(t))$$

hence $(\Delta E_{sy})_i < 0$.

(ii) For $|x_i(t) - x_i(t-2)| = 2$ we may have:

$t-2$	$t-1$	t
a	$a+1$	$a+2$
a	$a-1$	$a-2$

For the first case (the other is analogous) we get:

$$\sum_{j \in V_i} \delta(a+2, x_j(t)) - \delta^*(a+2, x_i(t)) > \sum_{j \in V_i} \delta(a, x_j(t)) - \delta^*(a, x_i(t))$$

so $(\Delta E_{sy})_i < 0$.

Then we conclude that $x(t) \neq x(t-2) \Longrightarrow E_{sy}(x(t)) < E_{sy}(x(t-1))$. ■

As in previous results, we may characterize the steady state directly from the above theorem:

Corollary 7.2. A sequential update admits only fixed points, while a synchronous update admits either fixed points or two-cycles. ■

An example of (7.20) is the median smoothing function:

$$g_i(x) = \begin{cases} x_i - 1 & \text{if} & |\{j \in V_i / x_j < x_i\}| < \frac{|V_k|}{2} \\ x_i + 1 & \text{if} & |\{j \in V_i / x_j > x_i\}| > \frac{|V_k|}{2} \\ x_i & & \text{otherwise} \end{cases} \qquad (7.21)$$

This rule belongs to the class defined in (7.20). In fact, by taking:

$$\delta(u, v) = \min(u, v); \quad \delta^*(u, v) = uv; \quad b_i = \frac{|V_i|}{2}, \quad \text{we have:}$$

Lemma 7.3. $\forall x \in Q^n$, $\forall i = 1, ..., n$, $g_i(x) = f_i(x)$ where:

$$f_i(x) = \begin{cases} x_i - 1 & \text{if} & \sum_{j \in V_i} \min(x_i - 1, x_j) - \frac{|V_i|}{2}(x_i - 1) > \\ & & \sum_{j \in V_i} \min(s, x_j) - \frac{|V_i|}{2}s \quad \text{for } s = x_i, x_i + 1 \\ x_i + 1 & \text{if} & \sum_{j \in V_i} \min(x_i + 1, x_j) - \frac{|V_i|}{2}(x_i + 1) > \\ & & \sum_{j \in V_i} \min(s, x_j) - \frac{|V_i|}{2}s \quad \text{for } s = x_i, x_i - 1 \\ x_i & & \text{otherwise} \end{cases} \qquad (7.22)$$

Proof. Given $k \in \{1, ..., n\}$, equality $f_k(x) = x_k - 1$ implies

$$\sum_{j \in V_k} (\min(x_k - 1, x_j) - \min(s, x_j)) + \frac{|V_k|}{2} > 0 \text{ for } s = x_k, \ x_k + 1$$

Hence for $s = x_k$ and since $x_k > x_j$ we obtain:

$$\min(x_k - 1, x_j) - \min(x_k, x_j) = x_j - x_j = 0$$

so

$$\sum_{\substack{j \in V_k \\ x_j \geq x_k}} \{x_k - 1 - x_k\} + \frac{|V_k|}{2} > 0$$

We deduce:

$$|\{j \in V_k / x_j \geq x_k\}| < \frac{|V_k|}{2} \quad \text{which implies} \quad |\{j \in V_k / x_j < x_k\}| > \frac{|V_k|}{2},$$

so $g_k(x) = x_k - 1$.

Now, if $f_k(x) = x_k + 1$ we get for $s = x_k$:

$$\sum_{j \in V_k} (\min(x_k + 1, x_j) - \min(x_k, x_j)) - \frac{|V_k|}{2} > 0$$

$$= \sum_{\substack{j \in V_k \\ x_k < x_j}} \{x_k + 1 - x_k\} - \frac{|V_k|}{2} > 0$$

so $\qquad |\{j \in V_k / x_j > x_k\}| > \dfrac{|V_k|}{2}$

then $g_k(x) = x_k + 1$.

If $f_k(x) = x_k$, from previous cases and for $s = x_k$ we get:

$$\sum_{j \in V_k} (\min(x_k - 1, x_j) - \min(x_k, x_j)) + \frac{|V_k|}{2} \leq 0$$

and

$$\sum_{j \in V_k} (\min(x_k + 1, x_j) - \min(x_k, x_j)) - \frac{|V_k|}{2} \leq 0$$

Hence

$$-|\{j \in V_k / x_j \geq x_k\}| + \frac{|V_k|}{2} \leq 0$$

and

$$|\{j \in V_k / x_j > x_k\}| - \frac{|V_k|}{2} \leq 0$$

This means $|\{j \in V_k / x_j \geq x_k\}| \geq \frac{|V_k|}{2}$ and $|\{j \in V_k / x_j > x_k\}| \leq \frac{|V_k|}{2}$, so

$|\{j \in V_k / x_j < x_k\}| \leq \dfrac{|V_k|}{2}$ and $|\{j \in V_k / x_j > x_k\}| \leq \dfrac{|V_k|}{2}$, hence $g_k(x) = x_k$.

Conversely, let us suppose that $g_k(x) = x_k - 1$. We deduce:

$$|\{j \in V_k / x_j < x_k\}| > \frac{|V_k|}{2}$$

If $s = x_k$ we have:

$$\Psi = \sum_{j \in V_k} [\min(x_k - 1, x_j) - \min(x_k, x_j)] + \frac{|V_k|}{2}$$

$$= -|\{j \in V_k / x_j \geq x_k\}| + \frac{|V_k|}{2} > 0$$

Now, if $s = x_k + 1$ we get:

$$\Psi = \sum_{j \in V_k} [\min(x_k - 1, x_j) - \min(x_k + 1, x_j)] + |V_k|$$

$$= \sum_{\substack{j \in V_k \\ x_j = x_k}} \{x_k - 1 - x_k\} + \sum_{\substack{j \in V_k \\ x_j > x_k}} \{x_k - 1 - x_k - 1\} + |V_k|$$

$$= -|\{j \in V_k / x_j = x_k\}| - 2|\{j \in V_k / x_j > x_k\}| + |V_k|$$

$$= -|\{j \in V_k / x_j \geq x_k\}| + \frac{|V_k|}{2} - |\{j \in V_k / x_j > x_k\}| + \frac{|V_k|}{2}$$

Since $|\{j \in V_k | x_j < x_k\}| > \frac{|V_k|}{2}$ we obtain:

$$\Psi > \frac{|V_k|}{2} - |\{j \in V_k / x_j > x_k\}| > 0$$

and we conclude, for $s = x_k, x_k + 1$:

$$\sum_{j \in V_k} \min(x_k - 1, x_j) - \frac{|V_k|}{2}(x_k - 1) > \sum_{j \in V_k} \min(s, x_j) - \frac{|V_k|}{2} s; \text{ for } s = x_k, x_k + 1$$

that is to say, $f_k(x) = x_k - 1$.

If $g_k(x) = x_k + 1$ we have $|\{j \in V_k / x_j > x_k\}| > \frac{|V_k|}{2}$. For $s = x_k$ we get:

$$\sum_{j \in V_k} (\min(x_k + 1, x_j) - \min(x_k, x_j)) - \frac{|V_k|}{2} = |\{j \in V_k / x_j > x_k\}| - \frac{|V_k|}{2} > 0$$

And for $s = x_k - 1$:

$$\sum_{j \in V_k} (\min(x_k + 1, x_j) - \min(x_k - 1, x_j)) - |V_k|$$

$$= |\{j \in V_k / x_j \geq x_k\}| + |\{j \in V_k / x_j > x_k\}| - |V_k|$$

$$> |\{j \in V_k / x_j > x_k\}| - \frac{|V_k|}{2} > 0$$

We conclude $f_k(x) = x_k + 1$.

Finally, if $g_k(x) = x_k$: $|\{j \in V_k / x_j < x_k\}| \leq \frac{|V_k|}{2}$ and $|\{j \in V_k / x_j > x_k\}| \leq \frac{|V_k|}{2}$. We have to prove that $f_k(x) = x_k$. It suffices to show that for the two inequalities in the definition of the function f_k (see (7.22)), at least one of the two values of s is such that the inequalities do not hold.

By taking the first inequality of (7.22) for $s = x_k$ we get:

$$\sum_{j \in V_k} \left(\min(x_k - 1, x_j) - \min(x_k, x_j)\right) + \frac{|V_k|}{2}$$

$$= -|\{j \in V_k / x_j \geq x_k\}| + \frac{|V_k|}{2}$$

$$= -|\{j \in V_k / x_j = x_k\}| - |\{j \in V_k / x_j > x_k\}| + \frac{|V_k|}{2}$$

$$\leq -|\{j \in V_k / x_j = x_k\}| \leq 0$$

then $f_k(x) \neq x_k + 1$.

For the second inequality and still for $s = x_k$, we get:

$$\sum_{j \in V_k} \left(\min(x_k + 1, x_j) - \min(x_k, x_j)\right) - \frac{|V_k|}{2} = |\{j \in V_k | x_j > x_k\}| - \frac{|V_k|}{2} \leq 0$$

then $f_k(x) \neq x_k - 1$ and we conclude $f_k(x) = x_k$, which proves the lemma. ∎

Function (7.21) may be seen as a "local median strategy" for smoothing digital images with $q = |Q|$ gray levels. Clearly, from previous results the Lyapunov operator for the synchronous iteration is:

$$E_{sy}(x(t)) = - \sum_{(i,j) \in V} \min(x_i(t), x_j(t-1)) + \sum_{i \in I} \frac{|V_i|}{2}(x_i(t) + x_i(t-1))$$

Also, $\{E_{sy}(x(t))\}_{t \geq 1}$ is strictly decreasing in the transient phase and therefore the synchronous dynamics converges to fixed points and/or two cycles.

7.4.3. The Phase Unwrapping Algorithm. The phase unwrapping algorithm computes the argument (phase) of a complex function. In practice this procedure is used in signal proccesing, solid-state physics, adaptative optics, etc. More information about the problem may be seen in [GMR,OR].

Since phase unwrapping is roughly a process of removing discontinuities by local neighborhood test and corrections, a local cooperative model was developed in [GMR] based on a "strength-of-vote" rule. The algorithm is the following:

On a finite non-oriented graph, each node initially contains a sample of the phase principal value and all nodes are updated synchronously accordingly to the following rule:

$$x_i(t+1) = \begin{cases} x_i(t) - 1 & \text{if} \quad s_i(t) < 0 \\ x_i(t) & \text{if} \quad x_i(t) = x_j(t) \quad \forall j \in V_i \\ x_i(t) + 1 & \text{otherwise} \end{cases} \tag{7.23}$$

where

$$s_i(t) = \sum_{j \in V_i} x_j(t) - d_i x_i(t), \quad d_i = |V_i| \text{ and } x_j(t) \in Q \subset \{-M, ..., +M\} \subseteq \mathbb{Z}.$$

We may interpret rule (7.23) as a local average: each site tries to be in the direction of its local average.

It is easy to see that $x(0) \in \{-M, ..., +M\}$ implies $|x_i(t)| \le M$ for any $t \ge 0$. Hence, given an initial condition in $\mathbb{Z}$, the iteration remains in the finite set $Q = \{- \max_i |x_i(0)| = -M, ..., \max_i |x_i(0)| = M\}$. For instance in a 1-dimensional lattice take $x(0) \in \{0, ..., 9\}^9$. For $1 \le i \le 8$ we have:

$$x_i(t+1) = \begin{cases} x_i(t) - 1 & \text{if} \quad x_{i+1}(t) + x_{i-1}(t) - 2x_i(t) < 0 \\ x_i(t) & \text{if} \quad x_i(t) = x_{i+1}(t) = x_{i-1}(t) \\ x_i(t) + 1 & \text{otherwise} \end{cases}$$

and

$$x_0(t+1) = \begin{cases} x_0(t) - 1 & \text{if} \quad x_1(t) - x_0(t) < 0 \\ x_0(t) & \text{if} \quad x_1(t) = x_0(t) \\ x_0(t) + 1 & \text{otherwise} \end{cases}$$

$$x_8(t+1) = \begin{cases} x_8(t) - 1 & \text{if} \quad x_7(t) - x_8(t) < 0 \\ x_8(t) & \text{if} \quad x_7(t) = x_8(t) \\ x_8(t) + 1 & \text{otherwise} \end{cases}$$

The dynamics of $x(0) = (8, 9, 9, 3, 0, 5, 5, 9, 8)$ is exhibited in Figure 7.15.

0	1	2	3	4	5	6	7	8	site number
8	9	9	3	0	5	5	9	8	$t = 0$
9	8	8	4	1	4	6	8	9	1
8	9	7	5	2	3	7	7	8	2
9	8	8	4	3	4	6	8	7	3
8	9	7	5	4	5	7	7	8	4
9	8	8	6	5	6	6	8	7	5
8	9	7	7	6	5	7	7	8	6
9	8	8	6	7	6	6	8	7	7
8	9	7	7	6	7	7	7	8	8	transient
9	8	8	6	7	6	7	8	7	9	$20 = 2n$
8	9	7	7	6	7	8	7	8	10
9	8	8	6	7	8	7	8	7	11
8	9	7	7	8	7	8	7	8	12
9	8	8	8	7	8	7	8	7	13
8	9	8	7	8	7	8	7	8	14
9	8	9	8	7	8	7	8	7	15
8	9	8	9	8	7	8	7	8	16
9	8	9	8	9	8	7	8	7	17
8	9	8	9	8	9	8	7	8	18
9	8	9	8	9	8	9	8	7	19
8	9	8	9	8	9	8	9	8	20	two-cycle
9	8	9	8	9	8	9	8	9	

Figure 7.15. Synchronous dynamics of the $1 - D$ phase-unwrapping algorithm.

The theoretical aspects of this algorithm have been studied essentially in [G6,GOd,OR,P]. The first formal analysis was given in [OR], where it was established that the quadratic operator:

$$H(x(t)) = - < x(t), Ax(t-1) >$$
$$\text{with} \quad a_{ij} = \begin{cases} 1 & \text{if} \quad j \in V_i \quad (\text{i.e. } (i,j) \in V) \\ -d_i = -|V_i| & \text{if } i = j \\ 0 & \text{otherwise} \end{cases} \qquad (7.24)$$

is non-increasing.

Unfortunately $\Delta_t H$ may vanish in the transient time (i.e. for t such that $x(t+2) \neq x(t)$), hence no bounds for transient time may be given from it. Nevertheless, the previous expression is powerful enough to characterize the steady state: for non oriented graphs the algorithm converges to fixed points or two cycles.

In order to give bounds for convergence a more sophisticated analysis is needed, in fact rule (7.23) may be studied in the framework of Potts models by taking a slightly more general Hamiltonian:

$$H(x) = -\frac{1}{2} \sum_{(i,j)\in V} a_{ij} \delta^1(x_i, x_j) - \frac{1}{2} \sum_{i\in I} \sum_{j\in V_i \cup \{i\}} \delta^2(x_i, x_j) + \sum_{i\in I} \delta^*(b_i, x_i)$$

where $A = (a_{ij})$ is the symmetric matrix defined in (7.24) and:

$$\delta^1(u,v) = \alpha uv; \quad \alpha = 4 \max_i d_i + 5$$
$$\delta^2(u,v) = 2 \min(u,v)$$
$$\delta^*(u,v) = uv \text{ and } b_i = 1 \quad \forall i = 1, ..., n$$

We obtain the following result:

Theorem 7.5. [GOd]. For the synchronous iteration, $E_{sy} x(t)$, defined by:

$$E_{sy}(x(t)) = -\alpha \sum_{i=1}^{n} \sum_{j=1}^{n} a_{ij} x_i(t) x_j(t-1) - 2 \sum_{i=1}^{n} \sum_{j\in V_i \cup \{i\}} \min(x_i(t), x_j(t-1))$$
$$+ \sum_{i=1}^{n} (x_i(t) + x_i(t-1))$$

is a strictly decreasing Lyapunov functional.

Proof. Since A is symmetric and G is non-oriented:

$$(\Delta_t E_{sy})_i = E_{sy}(x(t+1)) - E_{sy}(x(t)) = \sum_{i\in I} (\Delta_t E_{sy})_i$$

where the i-th component is:

$$(\Delta_t E_{sy})_i = -\alpha(x_i(t+1) - x_i(t-1))s_i(t) + 2\min(x_i(t), x_i(t-1))$$
$$- 2\min(x_i(t+1), x_i(t)) + 2 \sum_{j\in V_i} (\min(x_j(t), x_i(t-1))$$
$$- \min(x_j(t), x_i(t+1))) + x_i(t+1) - x_i(t-1)$$

Clearly, if $x_i(t+1) = x_i(t-1)$ we get $(\Delta_t E_{sy})_i = 0$. Now, if $x_i(t+1) \neq x_i(t-1)$ and $s_i(t) \neq 0$:

$$
\begin{aligned}
(\Delta_t E_{sy})_i \leq &- \alpha + 2(\min(x_i(t), x_i(t-1)) - \min(x_i(t+1), x_i(t))) \\
&+ 2 \sum_{j \in V_i} (\min(x_j(t), x_i(t-1)) - \min(x_j(t), x_i(t+1))) \\
&+ x_i(t+1) - x_i(t-1) \\
\leq &- \alpha + 2 + 2 \sum_{j \in V_i} 2 + 2 \\
\leq &- \alpha + 2 + 4d_i + 2 = -4 \max_i d_i - 5 + 4 + 4d_i \leq -1
\end{aligned}
$$

On the other hand, if $x_i(t+1) \neq x_i(t-1)$ and $s_i(t) = 0$, two cases occur:

1. $\forall j \in V_i$: $x_j(t) = x_i(t)$. Hence $x_i(t+1) = x_i(t)$, therefore:

$$
\begin{aligned}
(\Delta_t E_{sy})_i = &2(\min(x_i(t), x_i(t-1)) - x_i(t)) \\
&+ 2 \sum_{j \in V_i} (\min(x_i(t), x_i(t-1)) - x_i(t)) + x_i(t) - x_i(t-1)
\end{aligned}
$$

Since $x_i(t+1) \neq x_i(t-1)$ and $x_i(t+1) = x_i(t)$ then $x_i(t-1) = x_i(t) - 1$ or $x_i(t-1) = x_i(t) + 1$.

For $x_i(t-1) = x_i(t) - 1$ we get:

$$
(\Delta_t E_{sy})_i = -2 + 2 \sum_{j \in V_i} (-1) + 1 = -1 - 2 \sum_{j \in V_i} 1 = -1 - 2d_i < 0
$$

for $x_i(t-1) = x_i(t) + 1$ we obtain:

$$
(\Lambda_t E_{sy})_i = x_i(t) - x_i(t) - 1 = -1 < 0.
$$

2. There exists $k \in V_i$ such that $x_k(t) \neq x_i(t)$, hence $x_i(t+1) = x_i(t) + 1$. Since $s_i(t) = 0$ and $x_k(t) \neq x_i(t)$ it is easy to see that there exist $l \in V_i$ such that $x_l(t) > x_i(t)$. We have to analyze two situations:

$$
\begin{array}{ccc}
t-1 & t & t+1 \\
x_i(t) & x_i(t) & x_i(t) + 1 \\
x_i(t) - 1 & x_i(t) & x_i(t) + 1
\end{array}
$$

In the first case any $j \in V_i$ satisfies:

$$\min(x_j(t), x_i(t-1)) - \min(x_j(t), x_i(t+1)) =$$

$$\min(x_j(t), x_i(t)) - \min(x_j(t), x_i(t)+1) = \begin{cases} 0 & \text{if } x_j(t) \leq x_i(t) \\ -1 & \text{if } x_j(t) > x_i(t) \end{cases}$$

In the second case any $j \in V_i$ satisfies:

$$\min(x_j(t), x_i(t-1)) - \min(x_j(t), x_i(t+1)) =$$

$$\min(x_j(t), x_i(t)-1) - \min(x_j(t), x_i(t)+1) = \begin{cases} 0 & \text{if } x_j(t) < x_i(t) \\ -1 & \text{if } x_j(t) = x_i(t) \\ -2 & \text{if } x_j(t) > x_i(t) \end{cases}$$

Hence, since there exists $l \in V_i$ such that $x_l(t) > x_i(t)$, we conclude that in both cases the following inequality holds:

$$\sum_{j \in V_i} \{\min(x_j(t), x_i(t-1)) - \min(x_j(t), x_i(t+1))\} \leq -1$$

Then in both cases we get:

$$(\Delta_t E_{sy})_i = 2\{\min(x_i(t), x_i(t-1)) - \min(x_i(t+1), x_i(t))\}$$

$$+ 2 \sum_{j \in V_i} \{\min(x_j(t), x_i(t-1)) - \min(x_j(t), x_i(t+1))\}$$

$$+ x_i(t+1) - x_i(t-1) \leq -1 < 0$$

This proves the theorem. ∎

Corollary 7.3. [GOd,OR] The phase unwrapping algorithm converges to fixed points or two-cycles.

Proof. Direct from the previous theorem. ∎

For the phase unwrapping algorithm we may use the previous Lyapunov functional to bound the transient time. Since $Q = \{-M, ..., +M\}$ is a finite set and E is strictly decreasing, it suffices to give bounds of E in Q and to calculate the minimum $|\Delta E|$ in the transient phase. Given $x(0) \in Q^n$, $-M \leq x_i(0) \leq M$, we have:

Theorem 7.6. [GOd]. Let G be a connected non-oriented graph and $x(0) \in Q^n$. The transient time for synchronous update of iteration (7.23) is bounded by:

$$\tau \leq 8\alpha Me + 16Me + 10M - 4 \tag{7.25}$$

where e is the number of edges in G and $M = \max_i |x_i(0)|$. ∎

Before proving this theorem let us show the following result:

Lemma 7.4. Under the above conditions:

$$\begin{align}
&\text{(i)} \quad |s_i(t)| \leq 2d_i \;\Rightarrow\; |s_i(t+1)| \leq 2d_i \\
&\text{(ii)} \quad |s_i(t)| > 2d_i \;\Rightarrow\; |x_i(t)| < M - 2
\end{align}$$

Proof. (i) If $0 < s_i(t) \leq 2d_i$ we have $x_i(t+1) = x_i(t) + 1$, so

$$-2d_i < s_i(t) - 2d_i \leq s_i(t+1) = \sum_{j \in V_i} x_j(t+1) - d_i(x_i(t)+1) \leq s_i(t) \leq 2d_i$$

Then $|s_i(t+1)| \leq 2d_i$.

For $-2d_i \leq s_i(t) < 0$ we have $x_i(t+1) = x_i(t) - 1$, so:

$$-2d_i \leq s_i(t) \leq s_i(t+1) \leq s_i(t) + 2d_i < 2d_i$$

Then $|s_i(t+1)| \leq 2d_i$.

Now, if $s_i(t) = 0$ two cases arise. The first one $\forall j \in V_i$: $x_i(t) = x_j(t)$, which implies $x_i(t+1) = x_i(t)$. Then:

$$-d_i \leq s_i(t+1) \leq d_i \;\Rightarrow\; |s_i(t+1)| \leq 2d_i$$

In the other case, there exists $j \in V_i$ such that $x_j(t) \neq x_i(t)$. This implies $x_i(t+1) = x_i(t) + 1$, hence:

$$-2d_i \leq s_i(t+1) \leq 0 \leq 2d_i \;\Rightarrow\; |s_i(t+1)| \leq 2d_i.$$

(ii) First let us suppose $s_i(t) > 2d_i$ so

$$\sum_{j \in V_i} x_j(t) - d_i x_i(t) > 2d_i$$

hence

$$d_i x_i(t) < \sum_{j \in V_i} x_j(t) - 2d_i \le M d_i - 2d_i$$

which implies $x_i(t) < M - 2$

If $s_i(t) < -2d_i$ we get $d_i x_i(t) > -M d_i + 2d_i$, so $x_i(t) > -M + 2$. We conclude

$$|s_i(t)| > 2d_i \Rightarrow |x_i(t)| < M - 2 \quad \blacksquare$$

Proof of Theorem 7.6. From part (i) of the above lemma, a site satisfying $|s_i(t^*)| \le 2d_i$ for some t^*, is a condition which remains in time; i.e.: $|s_i(t)| \le 2d_i$ $\forall t \ge t^*$.

On the other hand, if $|s_i(t)| > 2d_i$, it is easy to see that the maximum number of steps where this situation holds is $2(M-3)+1$. So:

$$x_i(0) = -M+3, \; x_i(1) = -M+4, ..., x_i(2(M-3)+1) = M+3 \quad \text{for } s_i(t) > 2d_i.$$

Then for any $t \ge 2M - 4$ $|s_i(t)| \le 2d_i$ $\forall i = 1, ..., n$. On the other hand

$$|E(t+1)| \le \alpha \sum_{i \in I} |x_i(t+1)||s_i(t)| + 2nM + 2M \sum_{i \in I} d_i + 2nM$$

so, for $t \ge 2M - 4$:

$$|E(t+1)| \le 2\alpha M \sum_{i \in I} d_i + 4nM + 2M \sum_{i \in I} d_i$$

Since $2e = \sum_{i \in I} d_i$ we get:

$$|E(t+1)| \le 4\alpha M e + 4M(n + e)$$

Since $|\Delta E| \ge 1$ we have:

$$\tau \le 8\alpha M e + 8M(n+e) + 2M - 4$$

G being connected $e \ge n - 1$ and we deduce:

$$\tau \le 8\alpha M e + 16M e + 10M - 4 \quad \blacksquare$$

Remark. Bound (7.25) may be written in a more compact form for large n, namely:

$$\tau \le 24\alpha M e$$

In fact, it is important to see that $\tau = 0(Me)$ and in a one-dimensional automaton of n sites: $e = n-1$ then $\tau = 0(Mn)$. In practice, in regular lattices the convergence is faster than our bound indicates (always $0(Mn)$) but no better bounds have been found. Using other tools in a similar problem (but a simpler one) narrow bounds where obtained [G6,P] but also of the order $0(Me)$ with better constants.

Easier to study are the iterations without the previous tie-break:

$$x_i(t+1) = x_i(t) \iff x_i(t) = x_j(t) \quad \forall j \in V_i$$

For instance:

$$x_i(t+1) = \begin{cases} x_i(t) - 1 & \text{if } s_i(t) < 0 \\ x_i(t) + 1 & \text{otherwise} \end{cases}$$

This one may be analyzed in the framework of cyclically monotone functions by generalizations of results presented in Chapter 5 [G6] and also directly from the definition of cyclically monotone function without using the notion of potential [P].

Typical patterns generated by the synchronous update of phase-unwrapping automaton $\mathcal{A} = (\mathbb{Z}_{100}, V_0 = \{-1,0,1\}, Q = \{0,...,15\}, f)$ are shown in Figures 7.16 and 7.17 for rule (7.23). Each integer in Q is coded in four bits as its binary representation. Black dots mean 1 and white dots 0.

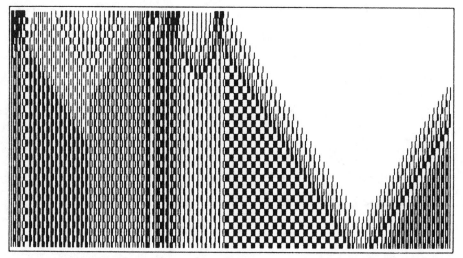

Figure 7.16. Synchronous dynamics of the $1 - D$ phase-unwrapping algorithm.

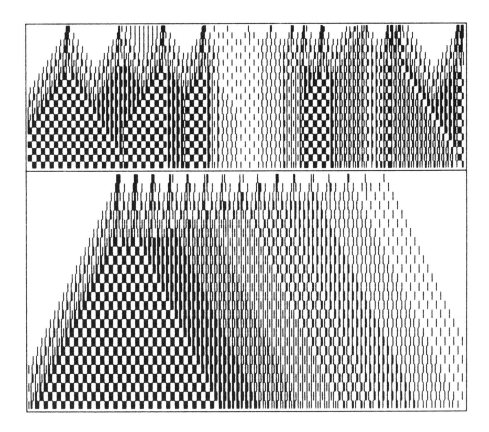

Figure 7.17. Synchronous dynamics of the $1 - D$ phase-unwrapping algorithm.

References

[G5]   Goles, E., *Potts Model and Automata Networks*, in Instabilities and
       in Nonequilibrium Structures II, E, Tirapegui and D. Villarroel eds., Kluwer,
       1989.

[G6]   Goles, E., *Local Graph Transformations Driven by Lyapunov Functionals*,
       Preprint Dept. Ing. Mat., E. de Ing., U. de Chile, 1988, submitted to Complex
       Systems.

[GMR] Ghiglia, D.C., G.A. Mastin and L.A. Romero, *Cellular Automata Method for
       Phase Unwrapping*, J. Optical Soc. America A, 4 (1987), 267-280.

[GOd] Goles, E., and A.M. Odlyzko, *Decreasing Energy Functions and Lengths of
       Transients for some C.A.*, Complex System, 2(5), 1988, 501-507

[GV] Goles, E. and G. Vichniac, *Attractors in Synchronous Networks of Multibit Threshold Automata*, preprint MIT Plasma Fusion Center, 1988, submitted to J. Phys.

[OR] Odlyzko, A.M. and D.J. Randall, *On the Periods of same Graph Transformations*, Complex Systems, 1, 1987, 203-210.

[P]   Poljak, S., *Transformations on Graphs and Convexity*, Complex Systems, 1, 1987, 1021-1033.

[PoV] Pomeau, Y. and G. Vichniac, *Extensions of Q2R: Potts Model and other Lattices*, J. Phys. A: Math. Gen. 21, 1988, 3297-3299.

[Pt] Potts, R.B. *Some Generalized Order-Disorder Transformations*, Proc. Camb. Phys. Soc., 48, 1952, 106-109.

[Wu] Wu, Y,F., *The Potts Model*, Review of Modern Physics, 54(1), 1982, 235-315.

REFERENCES

[A] Amari, S., *Homogeneous Nets of Neuron like Elements*, Biol. Cybern., 17, 1975, 211-220.

[AC] Albert, J., K. Culik, *A Simple Universal Cellular Automata and its one-way and Totalistic Version*, Complex Systems, 1(1), 1987, 1-16.

[AFS] Atlan, H., F. Fogelman-Soulie, J. Salomon, G. Weisbuch, *Random Boolean Networks*, Cybernetics and Systems, 12, 1981, 103.

[AGS] Amit, J.D., H. Gutfreund, M. Sompolinsky, *Spin-Glass Models of Neural Networks*, Phys. Review A, 32(2), 1985, 1007-1018.

[An] Anderson, J.A., *Cognitive and Phychological Computation with Neural Models*, IEEE Transactions on Systems, Man and Cybernetics, SMC-13, 1983, 799-815.

[AR] Allouche, J.P., Ch. Reder, *Oscillations Spatio-Temporelles Engendrees par un Automate Cellulaire*, Disc. Applied Maths., 1984, 215-254.

[B] Beurle, R.L., *Storage and Manipulation of Information in the Brain*, J. Inst. Elec. Eng., London, 1959, 75-82.

[Ba] Barahona, F., *Application de l'Optimisation Combinatoire à Certains Modeles de Verres de Spin*, Complexite et Simulations, Tesis, IMAG, Grenoble, France, 1980.

[Bax] Baxter, R.J., *Exactly Solved Models in Statistical Mechanics*, Academic Press, 1982.

[BCG] Berlekamp, E.R., J.H. Conway, R.K. Guy, *Winning Ways*, Ac. Press, 1985, 2, Chapter 25.

[BFW] Bienstock, E., F. Fogelman, G. Weisbuch (eds), *Disordered Systems and Biological Organization*, Nato ASI Series F: Computer and Systems Sciences, Springer-Verlag, 20, 1986.

[BM] Bidaux, R., P. Manneville (eds), *Proc. Workshop "Cellular Automata and Modeling of Complex Physical Systems"*, Les Houches, February 1989, to appear in Springer-Verlag.

[Br] Brascamp, H.J., *Equilibrium States for a One Dimensional Lattice Gas*, Comm. Math. Phys, 21, 1971, 56-70.

[C1] Caianiello, E.R., *Decision Equations and Reverberations*, Kybernetik, 3(2), 1966.

[C2] Caianiello, E.R., *Outline of a Theory of Thought-Processes and Thinking Machines*, J. Theor. Biol., 2, 1961, 204-235.

[CG] Cosnard, M., E. Goles, *Dynamique d'un Automate à Mémoire Modélisant le Fonctionnement d'un Neurone*, C.R. Acad. Sc., 299(10), Série I, 1984, 459-461.

[CMG] Cosnard, M., D. Moumida, E. Goles, T. de Saint Pierre. *Dynamical Behaviour of a Neural Automata with Memory*, Complex Systems, 2, 1988, 161-176.

[Co] Cood, E.F., *Cellular Automata*, Acad. Press, 1988.

237

[CP] Cailliez, F., J.P. Pages, *Introduction a l'Analyse des Données*, Smash, 1976.

[CY] Culik, K., S. Yu, *Undecidability of C.A. Classification Schemes*, Complex Systems, 2(2), 1988, 177-190.

 [D] Derrida, B., *Dynamics of Automata, Spin Glasses and Neural Network Models*, Lectures given at the school, Preprint, Service de Physique Théorique, CEN-Saclay, France, 1987.

[DGT] Demongeot, J., E. Goles, M. Tchuente (eds), *Dynamical Systems and Cellular Automata*, Acad. Press, 1985.

[DKT] Dobrushin, R.L., V.I. Kryukov, A.L. Toom, *Locally Interacting Systems and their Application in Biology*, Lecture Notes in Mathematics, N°653, 1978.

[DP] Derrida, B., Y. Pomeau, *Random Networks of Automata: A simple Annealed Approximation*, Europhysics Letters, 1(2), 1986, 45-49.

[DW] Derrida, B., G. Weisbuch, *Evolution of Overlaps between Configurations in Random Boolean Networks*, J. Physique, 47, 1986, 1297-1303.

 [F] Fogelman-Soulie, F., *Contribution a une Théorie du Calcul sur Reseaux*, Thesis, IMAG, Grenoble, 1985.

[FC] Farley, B., W.A. Clark, *Activity in Neurons-Like Elements*, in Proc. Fourth London Conference on Information Theory, C. Cherry (ed), Butterworths, London, 1961, 242-251.

[FG1] Fogelman-Soulie, F., E. Goles, G. Weisbuch, *Specific Roles of the Different Boolean Mappings in Random Networks*, Bull. Math. Biol., 44(5), 1982, 715-730.

[FG2] Fogelman-Soulie, F., E. Goles, G. Weisbuch, *Transient length in Sequential Iteration of Threshold Functions*, Disc. App. Maths., 6, 1983, 95-98.

[FGM] Fogelman-Soulié, F., E. Goles, S. Martínez, C. Mejía, *Energy Functions in Neural Networks with Continuous Local Functions* 1988, submitted Complex Systems.

[FH] Frisch, U., B. Hasslacher, S. Orszag, S. Wolfram, *Proc. of "Workshop on Large Nonlinear Systems"*, 1986, Complex Systems, 1(4), 1987.

[FRT] Fogelman-Soulie, F., Y. Robert, M. Tchuente (eds), *Automata Networks in Computer Science; Theory and Applications*, Nonlinear Science Series, Manchester Univ. Press, 1987.

[FV] Fannes, M., A. Verbeure, *On Solvable Models in Classical Lattice Systems*, Comm. Math. Phys. 96, 1984, 115-124.

[FW] Fogelman-Soulie, F., G. Weisbuch, *Random Iterations of Threshold Networks and Associative Memory*, SIAM J. on Computing, 16, 1987, 203-220.

[G1] Goles, E., *Dynamics on Positive Automata*, Theoret. Comput. Sci., 41, 1985, 19-31.

[G2] Goles, E., *Positive Automata Networks*, in Disordered Systems and Biological Organization, E. Bienenstock, F. Fogelman-Soulie, G. Weisbuch (eds), NATO ASI, Series, F20, Springer Verlag, 1986, 101-112.

[G3] Goles, E., *Comportement Dynamique de Reseaux d'Automates*, Thesis, IMAG, Grenoble, 1985.

[G4] Goles, E., *Sequential Iterations of Threshold Functions*, in Numerical Methods in the Study of Critical Phenomena, Delladora et al (eds), Springer-Verlag, Series in Sygernetics, 1981, 64-70.

[G5] Goles, E., *Potts Model and Automata Networks*, in Instabilities and Nonequilibrium Structures II, E, Tirapegui and D. Villarroel (eds), Kluwer, 1989.

[G6] Goles, E., *Local Graph Transformations Driven by Lyapunov Functionals*, Preprint Dept. Ing. Mat., E. de Ing., U. de Chile, 1988, submitted to Complex Systems.

[G7] Goles. E., *Fixed Point Behaviour of Threshold Functions on a Finite Set*, SIAM J. on Alg. and Disc. Meths., 3(4), 1982, 529-531.

[G8] Goles, E., *Lyapunov Functions Associated to Automata Networks*, in Automata Networks in Computer Science, F. Fogelman, Y. Robert, M. Tchuente (eds), Manchester University Press, 1987, 58-81.

[G9] Goles. E., *Dynamical Behaviour of Neural Networks*, SIAM J. Disc. Alg. Meth., 6, 1985, 749-754.

[G10] Goles, E., *Antisymmetrical Neural Networks*, Disc. App. Math., 13, 1986, 97-100.

[Ga1] Galperin, G.A., *One-dimensional Automata Networks with Monotonic Local Interactions*, Problemy Peredachi Informatsii, 12(4), 1976, 74-87.

[Ga2] Galperin, G.A., *One-Dimensional Monotonic Tesselations with Memory* in Locally Interacting Systems and their Application in Biology, R.L. Dobrushin et al (eds), Lecture Notes in Mathematics, N°653, 1978, 56-71.

[GFP] Goles, E., F. Fogelman-Soulie, D. Pellegrin, *The Energy as a Tool for the Study of Threshold Networks*, Disc. App. Math., 12, 1985, 261-277.

[GGH] Greenberg, J.M., C. Greene, S.P. Hastings *A Combinatorial Problem Arising in the Study of Reaction-Diffusion Equations*, SIAM J. Algebraic and Discrete Meths, 1, 1980, 34-42.

[GHM] Goles, E., G. Hernández, M. Matamala., *Dynamical Neural Schema for Quadratic Discrete Optimization Problems*, Neural Networks, 1, Supplement 1, 1988, 96.

[GM1] Goles, E., S. Martínez, *A Short Proof on the Cyclic Behaviour of Multithreshold Symmetric Automata*, Information and Control, 51(2), 1981, 95-97.

[GM2] Goles, E., S. Martínez, *Properties of Positive Functions and the Dynamics of Associated Automata Networks*, Discrete Appl. Math. 18, 1987, 39-46.

[GM3] Goles, E., S. Martínez, *Lyapunov Functionals for Automata Networks defined by Cyclically Monotone Functions*, Preprint, 1987.

[GM4] Goles, E., S. Martínez, *The One-Site Distributions of Gibbs States on Bethe Lattice are Probability Vectors of Period ≤ 2 for a Nonlinear Transformation*, J. Stat. Physics, 52(1/2), 1988, 267-285.

[GM5] Goles, E., S. Martínez (eds), *Proc. Congrés Franco-Chilien en Math. Appliquées*, 1986, Revista de Matemáticas Aplicadas, 9(2), 1988.

[GM6] Goles, E., S. Martínez, *Exponential Transient Classes of Symmetric Neural Networks for Synchronous and Sequential Updating*, Preprint, Dep. Ing. Mat., Esc. Ing., U. Chile, 1989.

[GMR] Ghiglia, D.C., G.A. Mastin and L.A. Romero, *Cellular Automata Method for Phase Unwrapping*, J. Optical Soc. America A, 4 (1987), 267-280.

[GO1] Goles, E., J. Olivos, *Compartement Iteratif des Fonctions à Multiseuil*, Information and Control, 45(3), 1980, 800-813.

[GO2] Goles, E., J., Olivos, *The Convergence of Symmetric Threshold Automata*, Inf. and Control, 51(2), 1981, 98-104.

[GO3] Goles, E., J. Olivos, *Comportement Pèriodique des Fonctions à Seuil Binaires et Applications*, Disc. App. Math., 3, 1981, 95-105.

[GO4] Goles, E., J. Olivos, *Periodic Behaviour of Generalized Threshold Functions*, Disc. Maths., 30, 1980, 187-189.

[GOd] Goles, E., A.M. Odlyzko, *Decreasing Energy Functions and Lengths of Transients for some Lengths of Transients for Some Cellular Automata*, Complex Systems, 2(5), 1988, 501-507.

[Gr] Green, F., *NP-Complete Problems in Cellular Automata*, Complex Systems, 1(3), 1987, 453-474.

[GT1] Goles, E., M. Tchuente, *Iterative Behaviour of Generalized Majority Functions*, Math. Soc. Sci., 4, 1984.

[GT2] Goles, E., M. Tchuente, *Erasing Multithreshold Automata*, in Dynamical Systems and Cellular Automata, J. Demongeot, E. Goles, M. Tchuente (eds), Ac. Press, 1985, 47-56.

[GT3] Goles, E., M. Tchuente, *Iterative Behaviour of One-Dimensional Threshold Automata*, Disc. Appl. Maths. 8, 1984, 319-322.

[GV] Goles, E., G. Vichniac, *Attractors in Synchronous Networks of Multi Threshold Automata*, Preprint, MIT Plasma Fusion Center, 1988, submitted to J. Phys.

[HaL] Haken, A., M. Luby, *Steepest Descent can take Exponential Time for Symmetric Connection Networks*, Complex Systems 2, 1988, 191-196.

[He] Hedlund, G.A., *Endomorphism and Automorphisms of the Shift Dynamical System*, Math. System Theory, 3, 1969, 320-375.

[Ho1] Hopfield, J.J., *Neurons with Graded Response have Collective Computational Properties like those of two-state Neurons*, Proc. Nat. Acad, Sci, USA, 81, 1984, 3088-3092.

[Ho2] Hopfield, J.J., *Neural Networks and Physical Systems with Emergent Collective Computational Abilities*, Proc. Natl. Acad. Sci. USA., 79, 1982, 2554-2558.

[HoT] Hopfield, J.J., D.W. Tank, *Neural Computation of Decisions in Optimization Problems*, Biol. Cybernetics, 52, 1985, 141-152.

[HP1] Hardy, J., O. de Pazzis, Y. Pomeau, *Time Evolution of a Two-Dimensional Model System: Invariant States and Time Correlation Functions*, J. Math. Phy., 14, 1973, 174.

[HP2] Hardy, J., O. de Pazzis, Y. Pomeau, *Molecular Dynamics of a Classical Lattice Gas: Transport Properties and Time Correlation Functions*, Phys. Rev. A. 13, 1976, 1949.

[Hu] Hurd, L., *Formal Language Characterisation of Cellular Automaton Limit Sets*, Complex Systems, 1(1), 1987, 69-80.

[K] Kleene, S.C., *Representation of Events in Nerve Nets and Finite Automata* in Automata Studies, C.E. Shannon and J. McCarthy (eds), Annals of Mathematics Studies, 34, Princeton Univ. Press, 1956, 3-41.

[Ka] Kauffman, S.A., *Behaviour of Randomly Constructed Genetic Nets* in Towards a Theoretical Biology, C.H. Waddington (ed), 3, Edinburgh Univ. Press, 1970, 18-46.

[KGV] Kirkpatrick, S., C. Gelatt, M. Vecchi, *Optimization by Simulated Annealing*, Science, 220, 1983, 671-680.

[Ki1] Kitagawa, T., *Cell Space Approaches in Biomathematics*, Math. Biosciences, 19, 1974, 27-71.

[Ki2] Kitagawa, T., *Dynamical Systems and Operators Associated with a Single Neuronic Equation*, Math. Bios., 18, 1973.

[Ko] Kobuchi, Y., *Signal Propagation in 2-Dimensional Threshold Cellular Space*, J. of Math. Biol., 3, 1976, 297-312.

[KR] Krasnoselsky, M.A., Y.B. Rutitsky, *Convex Functions and Orlicz Spaces*, Industan Publ. Co., 1962.

[KS] Kindermann, R., J.L. Sneel, *Markov Random Fields and their Applications*, Series on Contemporary Mathematics, AMS, 1, 1980.

[L] Lind, D.A., *Applications of Ergodic Theory and Sofic Systems to Cellular Automata*, Physica, 10D, 1984, 36-44.

[Le] Legendre, M., *Analyse et Simulation de Reseaux d'Automates*, Thesis, IMAG, Grenoble, 1982.

[Li] Little, W.A. *Existence of Persistent States in the Brain*, Math. Bios., 19, 1974, 101.

[LiS] Little, W.A., G.L. Shaw, *Analytic Study of the Memory Storage Capacity of a Neural Network*, Math. Bios, 39, 1978, 281-290.

[LN] Lindgren, K., M. Nordahl, *Complexity Measures and Cellular Automata*, Complex Systems, 2(4), 409-440.

[M1] Martínez, S., *Lyapunov Functionals on Bethe Lattice*, in Proceedings Workshop on Disordered Systems, Bogotá, World Sc. Publ., 1989, 22-37.

[M2] Martínez, S., *Cylinder Distribution of Thermodynamic Limit of Bethe Lattice*, in Proceedings Symposium on Non Equilibrium Structures, Valparaíso, D. Reidel Publ. Co., 1989.

[M3] Martínez, S., *Relations among Discrete and Continuous Lyapunov Functionals for Automata Networks*, Preprint, 1989.

[M4] Martínez, S., *Chain-Symmetric Automata Networks*, Preprint, 1989.

[MaP] Marr, D., T. Poggio, *Cooperative Computation of Stereo-Disparity*, Science, 194.

[Mi] Minsky, M.L., *Computation: Finite and Infinite Machines*, Prentice-Hall, Series in Automatic Computation, 1967.

[MiP] Minsky, M., S. Papert, *Perceptrons, an Introduction to Computational Geometry*, MIT Press, 1969.

[Ml] Milnor, J., *On the Entropy Geometry of Cellular Automata*, Complex Systems, 2(3), 1988, 257-385.

[MP] McCulloch, W., W. Pitts, *A Logical Calculus of the Ideas Immanent in Nervous Activity*, Bull. Math. Biophysics, 5, 1943, 115-133.

[MPV] Mezard, M., G. Parisi, M.A. Virasoro (eds), *Spin Glass Theory and Beyond*, Lecture Notes in Physics, 9, World Scientific, 1987.

[NS] Nagumo, J., S. Sato, *On a Response Characteristic of a Mathematical Neuron Model*, Kybernetic, 3, 1972, 155-164.

[OR] Odlyzko, A.M. and D.J. Randall, *On the Periods of some Graph Transformations*, Complex Systems, 1, 1987, 203-210.

[P] Poljak, S., *Transformations on Graphs and Convexity*, Complex Systems, 1, 1987, 1021-1033.

[PA1] Peterson, C., J. Anderson, *A Mean Field Theory Learning Algorithm for Neural Networks*, Complex Systems, 1(5), 1987, 995-1019.

[PA2] Peterson, C., J. Anderson, *Neural Networks and NP-Complete Optimization Problems; A Performance Study on the Graph Bisection Problem*, Complex Systems, 2(1), 1988, 59-89.

[PE] Pham Dinh Tao, S. El Bernoussi, *Iterative Behaviour, Fixed Point of a Class of Monotone Operators. Application to Non-Symmetric Threshold Functions*, Disc. Maths., 70, 1988, 85-101.

[Pe] Peliti, L. (ed), *Disordered Systems and Biological Models*, Proc. of the Workshop and Disordered Systems and Biol. Modelling, Bogotá Colombia, 1987, World Scientific, CIF Series, 14, 1989.

[Per] Peretto, P., *Collective Properties of Neural Networks: A Statistical Physics Approach*, Biol. Cybern., 50, 1984, 51-62.

[Po] Pomeau, Y., *Invariant in Cellular Automata*, J. Phys., A17, 1984, L415-L418.

[PoV] Pomeau, Y. and G. Vichniac, *Extensions of Q2R: Potts Model and other Lattices*, J. Phys. A: Math. Gen., 21, 1988, 3297-3299.

[PS] Poljak, S., M. Sura, *On Periodical Behaviour in Society with Symmetric Influences*, Combinatorica, 3, 1983, 119-121.

[Pt] Potts, R.B. *Same Generalized Order-Disorder Transformations*, Proc. Camb. Phys. Soc., 48, 1952, 106-109.

[PT1] Poljak, S., D. Turzik, *On Pre-Periods of Discrete Influence Systems*, Disc. Appl. Maths., 13, 1986, 33-39.

[PT2] Poljak, S., D. Turzik, *On an Application of Convexity to Discrete Systems*, Disc. Appl. Math., 13, 1986, 27-32.

[Rck] Rockafellar, R.T., *Convex Analysis*, Princeton Univ. Press, Princeton, NJ., 1970.

[Ri] Richardson, D., *Tessellation with Local Transformation*, J. Comput. & Systems Sci., 6, 1972, 373-388.

[RM] Rumelhart, D.E., J.C. McClelland (eds), *Parallel and Distributed Processing: Explorations in the Miscrostructure of Cognition*, MIT Press, 1986.

[Ro] Robert, F., *Discrete Iterations. A Metric Study*, Springer Series in Computational Mathematics, Springer-Verlag, 1986.

[RoT] Robert, Y., M. Tchuente, *Connection-Graph and Iteration-Graph of Monotone Boolean Functions*, Disc. Appl. Maths., 11,1985, 245-253.

[Ru] Ruelle, D., *Thermodynamic Formalism*, Addison-Wesley, 1978.

[Sh1] Shingai, R., *Maximum Period of 2-Dimensional Uniform Neural Networks*, Inf. and Control, 41, 1979, 324-341.

[Sh2] Shingai, R., *The Maximum Period Realized in 1-D Uniform Neural Networks*, Trans. IECE, Japan, E61, 1978, 804-808.

[Sm] Smith, A.R., *Simple Computation-Universal Cellular Spaces*, J. ACM, 18(3), 1971, 339-353.

[T1] Tchuente, M., *Contribution a l'Etude des Methodes de Calcul pour des Systemes de Type Cooperatif*, Thesis IMAG, Grenoble, France, 1982.

[T2] Tchuente, M., *Sequential Iteration of Parallel Iteration*, Theor. Comp. Sci., 48, 1986, 135-144.

[T3] Tchuente, M., *Evolution de Certains Automates Cellulaires Uniformes Binaires à Seuil*, Res. Repp., SANG 265, IMAG, Grenoble, France, 1977.

[TfM] Toffoli, T., M. Margolus, *Cellular Automata Machines: A New Environment for Modeling*, MIT Press, 1987.

[To]   Toom, A.L., *Monotonic Binary Cellular Automata*, Problemy Peredaci Informacii, 12(1), 1976, 48-54.

[ToM] Toom, A.L., L.G. Mitynshin, *Two Results Regarding Noncomputability for Univariate Cellular Automata*, Problemy Peredaci Informacii, 12(2), 1976, 69-75.

[U]   Ulam S., *On Some Mathematical Problems Connected with Patterns of Growth of Figures* in Essays on Cellular Automata, A.W. Burks (ed), Univ. of Illinois Press, 1970, 219-243.

[V1]   Vichniac, G., *Simulating Physics with Cellular Automata*, Physica 10D, 1984, 96-116.

[V2]   Vichniac G., *Cellular Automata Models of Disordered and Organization* in Disordered Systems ans Biol. Org., E. Bienenstock et al (eds), NATO ASI Series F, 20, 1986, 3-19.

[VN1] Von Neumann, J., *Theory of Self-Reproducing Automata*, A.W. Burks (ed), Univ. of Illinois Press, 1966.

[VN2] Von Neumann, J., *The General and Logical Theory of Automata* in Hixon Synposium Proc., 1948 in J.N. Neumann Collected Works, A.H. Taub (ed), Pergamon Press, V,288-328, 1963.

[W1]   Wolfram, S., *Theory and Applications of Cellular Automata*, World Scientific, 1986.

[W2]   Wolfram, S., *Universality and Complexity in Cellular Automata*, Physica, 10D, 1984, 1-35.

[W3]   Wolfram, S., *Twenty Problems in the Theory of Cellular Automata*, Phys. Scripta, 9, 1985, 170.

[Wa]   Waksman, A., *A Model of Replication*, J.A.C.M., 16(1), 1966, 178-188.

[Wi]   Winograd, T., *A Simple Algorithm for Self-Reproduction*, MIT, Project MAC, Artificial Intelligence, Memo 198, 1970.

[Wu]   Wu, Y,F., *The Potts Model*, Review of Modern Physics, 54(1), 1982, 235-315.

AUTHOR AND SUBJECT INDEX